Stem Cell Biology and Regenerative Medicine

Volume 78

Our understanding of stem cells has grown rapidly over the last decade. While the apparently tremendous therapeutic potential of stem cells has not yet been realized, their routine use in regeneration and restoration of tissue and organ function is greatly anticipated. To this end, many investigators continue to push the boundaries in areas such as the reprogramming, the stem cell niche, nanotechnology, biomimetics and 3D bioprinting, to name just a few. The objective of the volumes in the Stem Cell Biology and Regenerative Medicine series is to capture and consolidate these developments in a timely way. Each volume is thought-provoking in identifying problems, offering solutions, and providing ideas to excite further innovation in the stem cell and regenerative medicine fields.

Ozge Sezin Somuncu
Editor

Organoids for Preclinical Disease Modeling and Personalized Medicine

Springer

Editor
Ozge Sezin Somuncu
Department of Radiation Oncology
Dana-Farber/Harvard Cancer Center
Boston, MA, USA

ISSN 2196-8985 ISSN 2196-8993 (electronic)
Stem Cell Biology and Regenerative Medicine
ISBN 978-3-032-10023-8 ISBN 978-3-032-10024-5 (eBook)
https://doi.org/10.1007/978-3-032-10024-5

This Springer imprint is published by the registered company Springer Nature Switzerland AG
The registered company address is: Gewerbestrasse 11, 6330 Cham, Switzerland

Preface

This book emerged in response to the rapid progress and increasing interest in the field of Organoids for Preclinical Disease Modeling, especially in the field of oncology. It brings together current research, innovative methods, and interdisciplinary perspectives to provide a thorough overview of recent developments.

The concept for this volume was thoroughly discussed, where contributors saw the benefit of consolidating their work into a unified collection. Each chapter has undergone peer review and careful revision to ensure quality and clarity, and we appreciate the reviewers for their constructive input.

This volume is intended for graduate students, researchers, clinicians, and professionals who are looking for an in-depth resource on clinical organoid research. It also aims to support ongoing work in the field and inspire future research directions.

We extend our sincere thanks to the contributing authors for their efforts, and to Springer Nature for their continued support in bringing this book to publication. I am personally grateful to Kursad Turksen and Lauren Pecorino for their endless scientific support throughout the years and to Alan D'Andrea for his mentorship in the organoid field. I dedicate this book to my mother, who is a breast cancer survivor, and to my family.

We hope this book serves as a helpful reference and sparks further exploration in this evolving area of study.

Boston, MA, USA
October 2025

Ozge Sezin Somuncu

Contents

Chapter 1
Preclinical Patient-Derived Tumor Organoids for Modeling Cancer

Ozge Sezin Somuncu

Abstract Over the last 10 years, patient-derived tumor organoids (PDTOs) have significantly expanded the range of preclinical models and gradually transformed the landscape of three-dimensional cell culture in cancer research. PDTOs can be efficiently generated from patient tumor samples and accurately reproduce the histological and molecular features of the original tumor. As a result, PDTOs represent powerful tools for oncology research, and their potential application in clinical settings holds great promise for advancing precision medicine in cancer treatment. In this chapter, we present a comprehensive overview of the techniques used for organoid growth, important elements for the preservation of tumor heterogeneity, the assays that can be implemented, the current organoid biobanks and extending to the discovery of novel therapeutic targets, preclinical testing of new anticancer agents, and advancements in precision medicine. Ongoing technological advancements and the development of autologous coculture systems for PDTOs aim to address current challenges and fully harness the potential of these models. Integrating PDTOs as routine tools in clinical oncology could pave the way for a new era of precision cancer care in the next decade.

Keywords Patient-derived tumor organoids · PDTOs · Personalized cancer treatment · Modeling cancer

1.1 Introduction

The development of effective cancer therapies relies heavily on accurate preclinical models that recapitulate the complexity of human tumors. Traditional two-dimensional (2D) cell cultures and xenograft models have provided foundational

O. S. Somuncu (✉)
Department of Radiation Oncology, Dana-Farber/Harvard Cancer Center, Boston, MA, USA
e-mail: ozges_somuncu@dfci.harvard.edu

O. S. Somuncu (ed.), *Organoids for Preclinical Disease Modeling and Personalized Medicine*, Stem Cell Biology and Regenerative Medicine 78,
https://doi.org/10.1007/978-3-032-10024-5_1

insights, yet they often fail to fully capture the heterogeneity, architecture, and microenvironmental interactions of actual patient tumors. In recent years, patient-derived tumor organoids (PDTOs) have emerged as a promising three-dimensional (3D) in vitro model that bridges this gap.

In the last decade, patient-derived tumor organoids (PDTOs) transformed precision oncology. PDTOs can be produced via tumor cells from biopsy samples, resection samples, or biological fluids such as ascites [1]. Mainly, cancer tissues are exposed to mechanical and enzymatic dissociation for obtaining isolated cells. Later on, these cells are embedded in an extracellular matrix (ECM) and are cultured with optimized media corresponding to specific tumor [1, 2]. It is important to generate organoids that recapitulate the original tumor's histology [2, 3] and genetic background [4] in addition to preserved clonal heterogeneity [5]. This chapter mainly focuses on the importance of PDTOs as their use continues to expand across various cancer types, and how they are redefining preclinical research by providing a more physiologically relevant and patient-specific model for cancer investigation.

1.2 Important Components of Organoid Growth

The main factor that is important for the organoid culture success is the obtainable initial material such as resection tissue versus biopsy and/or tumor cellularity. They greatly differentiate between different tumor types, and they establish the rate-limiting step. If the initiation of organoids is successful, the other factor for the growth rate of culture depends on whether the organoids accurately recapitulate features of the original tumor, which helps avoid bias in in vitro drug assays [6].

The ECM selection is essential for supporting organoid growth, and it provides the vital microenvironment for the self-organization. The most universally utilized commercial ECMs are the hydrogels originated from decellularized murine chondrosarcomas with or without growth factors (Matrigel, BME). They are mainly composed of laminin and collagen IV with a limitation of batch variabilities. Natural hydrogels can be listed as collagen hydrogels mixed with laminin, fibronectin, or hyaluronic acid [7].

In 2009, Hans Clevers' laboratory at the Hubrecht Institute in the Netherlands, a leader in this field, showed that a single adult intestinal stem cell expressing the LGR5 receptor, isolated from mice, could regenerate in culture, forming structures with cellular diversity and organization resembling the intestinal epithelium's crypts and villi [8]. These foundational findings have since been extended to multiple organs and adapted for culturing PDTOs, and later expanding to other cancer types [7]. As a result, research teams have successfully developed PDTOs from a range of cancers, including colorectal [9], lung [10], pancreatic [11], breast [12], ovarian [2, 13, 14], and prostate cancers [15].

During the growth of PDTOs, activation of two pathways is important: EGFR and the WNT pathways. EGFR pathway is essential for promoting cell proliferation and is achieved via EGF supplementation, while WNT is important for proliferation, adhesion, and cell differentiation, via stabilization of the β-catenin transcriptional co-factor [16, 17].

1.3 Preservation of Tumor Heterogeneity

The primary objective of the PDTO model is to develop an in vitro system that closely replicates the in vivo tumor microenvironment while preserving the original tumor characteristics, thereby reflecting human cancer physiology. PDTOs retain the genetic profiles and intra-tumoral heterogeneity of primary tumors. Additionally, well-established PDTOs biobanks can help uncover inter-tumoral heterogeneity across different cancer patient populations. To enhance PDTO cultivation, further research into optimizing matrices, soluble factors in the culture media, and co-culture systems that mimic the tumor microenvironment will provide valuable opportunities for maintaining the original tumor properties [18].

While patient-derived organoids (PDOs) preserve a significantly greater degree of heterogeneity compared to other models, those derived from a single tumor region have been found to be insufficient in fully capturing the clinical outcomes observed in actual patients. Thus, obtaining tissue samples from multiple regions (e.g., adjacent noncancerous and tumor tissue) is necessary to create a more accurate representation of the subpopulation's heterogeneity. The internal diversity within each patient-derived cancer organoid represents the complex makeup of tumor cell clones, which span from stem cell-like to more differentiated types, offering new insights into how these different clones interact and how they are affected by treatment [19]. A comprehensive genetic and epigenetic analysis of patients and their corresponding organoids will offer valuable insights and characteristics for assessing treatment responses [20].

1.4 Assays to Be Performed with PDTOs

1.4.1 Cell Viability Assays

Currently, the primary approach for evaluating organoid sensitivity to compounds involves cell viability assays, which are based on ATP measurements [21]. PDTOs can be treated with different treatments such as chemotherapy, radiotherapy, or targeted therapies, and their responses can be assessed via diverse viability tests, including CellTiter-Glo [2, 13], CellTiter Blue [22], MTS [23], and CCK-8 assays [24], as such are commonly utilized. Nevertheless, these methods have limitations, as it only provides data from the overall organoid population, lacking insight into the responses of specific cell subgroups and the temporal dynamics of drug effects. To overcome these challenges, advanced image-based high-throughput screening techniques and platforms have been developed. These approaches allow for the assessment of phenotypic changes in organoids and real-time evaluation of compound efficacy in organoid cultures [25].

1.4.2 Profiling Assays

A key advantage of using PDTOs for profiling is their ability to overcome challenges commonly faced when directly analyzing tumor tissues obtained from biopsies or surgical resections. These challenges include difficulties in detecting DNA, RNA, proteins, or metabolic features [3]. In many patient-derived samples, tumor cells may constitute only a small fraction, leading to potential contamination of the results by signals from non-tumor cells. For example, specific gene mutations in tumor cells can be masked by dominant signals from surrounding non-tumor cells [3]. Additionally, the quality of tumor samples may be compromised due to factors such as necrotic regions, which are prevalent in many tumors, or prior treatments like chemotherapy and radiation, which can affect the accuracy of profiling [26]. PDTOs help mitigate these issues by enriching viable tumor cells during in vitro culture, leading to a stronger representation of tumor-specific DNA mutations, RNA expression, and tumor-associated proteins. This enrichment enhances the reliability of biochemical profiling, making it more feasible to assess characteristics such as the metabolic status of PDTOs [27].

Organoid models represent a cutting-edge bioengineering method that captures the abnormal genetic alterations found in cancer patients, mirroring the genetic and functional traits of cancer cells. As a result, they have been utilized to create realistic models of tumor initiation, progression, and metastasis, and to evaluate the effectiveness of different treatments for various cancers, such as breast and ovarian cancer [28].

1.4.3 Immunofluorescence Assays

Immunofluorescence (IF) is an extensively used method to visualize target antigens with high precision and specificity for the precise identification and localization of proteins, glycans, and small molecules. It is a valuable technique for visualizing the spatial distribution of cell types within organoids and examining their interactions. Since organoids closely mimic the morphology of patient tumors, immunofluorescent staining and imaging provide deeper insights into structures that closely resemble those found in vivo [29]. Although this technique is well established in monolayer cell cultures, less is known about its use in three-dimensional models. Recently, novel methods for efficiently staining the PDTOs were established by different groups [29, 30].

The required use of extracellular matrix (ECM) gels in 3D cultures presents challenges, such as restricted antibody penetration and elevated background signals, while removing the ECM gel can disrupt organoid structure and lead to sample loss. These issues complicate efficient immune labeling [31]. Protocols such as whole-mount immunofluorescence staining of gel-embedded organoids seem to overcome these challenges [31]. While FFPE sections enable visualization of cells in the inner regions of organoids using histological techniques like hematoxylin and eosin (H&E), IF, and immunohistochemistry (IHC), they are limited to cross-sectional views and may distort organoid morphology due to the embedding process. In contrast, whole-mount staining allows organoids to be stained and examined in their intact form, preserving their structural features and enabling detailed imaging of protein localization. Originally used for staining entire tissues or organisms, such as zebrafish or mouse embryos, whole-mount techniques can be readily adapted for use with organoids [32].

1.4.4 Fluorescent-Based DNA Fiber Assays

Fluorescent-based DNA fiber assays are a type of biophysical technique developed for visualizing individual DNA replication fork (RF) dynamics at the molecular level. The method involves the labeling of newly synthesized DNA strands in vitro with nucleotide analogues, which are stained with fluorophore-conjugated antibodies. The results of this are fluorescently stained daughter strands that can be visualized using a variety of fluorescence imaging techniques, such as fluorescence microscopy or a specific scanner [2, 13, 33].

In 2018, Hill et al. showed that the stability of the replication fork may thus serve as an indicator of sensitivity or resistance to certain DNA repair therapies. In that study, Hill et al. generated short-term, patient-derived high-grade serous ovarian cancer (HGSOC) organoids (33 cultures from 22 patients), maintaining genetic and histological fidelity to parent tumors. These organoids were used to functionally test DNA repair capabilities—specifically homologous recombination (HR) proficiency (via RAD51 foci assays) and replication fork protection (via DNA fiber assays)—to predict response to drugs targeting DNA damage [33]. Later on, in a study by da Costa et al., our group used patient-derived, HGSOC organoids to study single-stranded DNA (ssDNA) gap accumulation as a biomarker (Fig. 1.1) for sensitivity to USP1 inhibitors (USP1—a deubiquitinating enzyme involved in DNA damage repair pathways, including the Fanconi anemia (FA) pathway and translesion synthesis (TLS)) [2]. Moreover, recent work by Ravindranathan et al. highlighted ssDNA gap accumulation in ovarian PDTOs is a versatile functional biomarker for predicting and potentially overcoming resistance to both PARP and PARG inhibitors. Especially promising perspective of this study was the ability of combined PARGi/PARPi therapy to sensitize resistant, *BRCA*-wildtype models through gap induction [13].

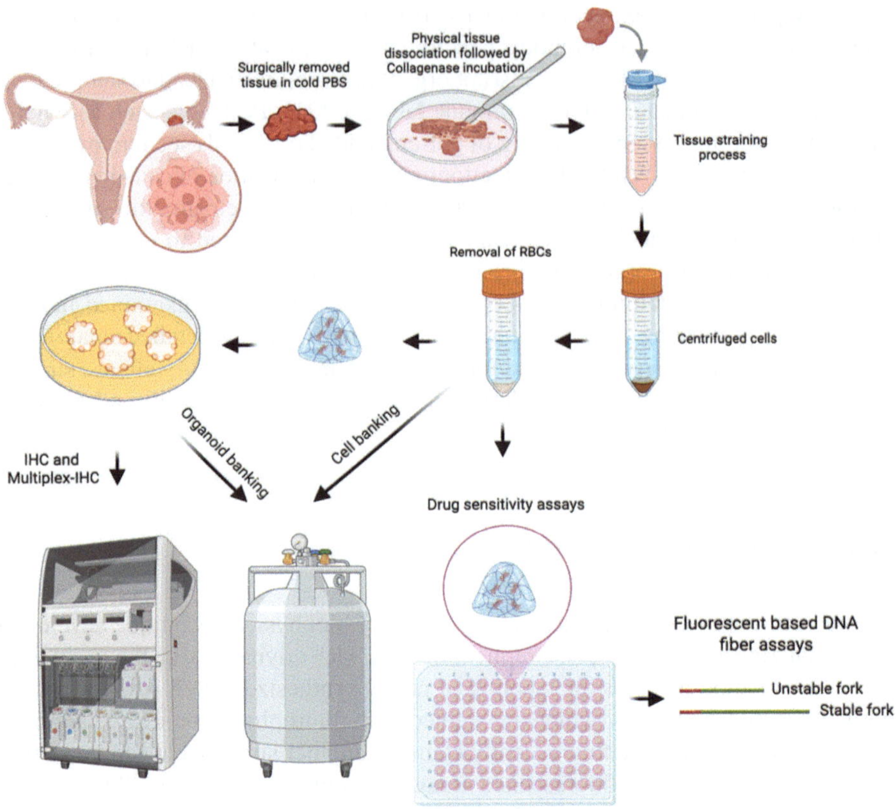

Fig. 1.1 Optimized ovarian tumor tissue processing for patient-derived tumor organoid (PDTO) production [2, 13, 14]. Ovarian cancer PDTOs can be used for downstream assays such as drug sensitivity assays, immunohistochemistry (IHC) analysis, and DNA fiber assays or can be banked for further analysis

1.4.5 Metabolic Assays

By using suitable platforms, organoids on-chip can be integrated with sensors to gather quantitative data and enable real-time kinetic analysis of PDTOs' responses to treatments. For example, these sensors could monitor oxygen concentrations [34] or metabolites like glucose and lactate [35], providing a more precise assessment of PDTO metabolism.

Using a PDTO biobank with defined metabolic profiles, researchers have identified biomarkers associated with pancreatic cancer recurrence. PDTOs from patients with early relapse showed elevated levels of certain Krebs cycle–derived oncometabolites. This metabolic feature may serve both as a marker of tumor aggressiveness and as a potential therapeutic target [36].

Optical metabolic imaging (OMI) is a highly sensitive method for detecting drug responses in organoids, and the responses observed in tumor organoids closely reflect how the primary tumor reacts to treatment. The entire workflow—from biopsy collection to identifying effective drugs—can be completed in under a week. Using organoid-based screening for anticancer drugs could help determine the most effective treatments for individual patients, maximizing tumor elimination while reducing the risk of unnecessary toxicity from ineffective therapies. When paired with OMI, a microscopy technique that detects metabolic changes at the cellular level, treatment response can be assessed based on drug-induced metabolic shifts [37, 38]. OMI also enables the analysis of tumor heterogeneity, identifying resistant subpopulations [39]. This type of organoid-based drug testing could be carried out in hospitals and clinics equipped with standard cell culture labs, making it accessible for broader clinical use.

In addition to fluorescence-based OMI, *matrix-assisted laser desorption/ionization mass spectrometry imaging (MALDI-MSI)* has been employed in multiple studies to map the distribution of anticancer drugs and their metabolites across various patient-derived tumor organoid (PDTO) models. These findings support the potential of combining OMI and MALDI-MSI as high-throughput platforms in PDTOs to identify personalized treatment options and to incorporate metabolic profiling into the current genomics-focused approach in precision oncology [40]. While advanced techniques such as metabolomics and ^{13}C tracer analysis are well established for measuring metabolite levels and metabolic pathway activity, they have rarely been applied in 3D organoid systems. Nevertheless, tumor organoid metabolomes have been characterized using NMR [41] and both targeted [42] and untargeted [43] liquid chromatography-mass spectrometry (LC-MS) approaches.

1.5 Patient-Derived Tumor Organoid Biobanks

Typically, tumor biobanks consist of normal and tumor tissues, along with occasionally matched blood samples. However, a major limitation of these biobanks is their non-renewable nature, making them suitable for only a single use [44]. Advancements in technology have enabled the creation of organoid models using 3D in vitro cell culture systems. Today, various companies and research institutions globally are gathering patient tissues to develop organoid biobanks, which is used for research in regenerative medicine, disease modeling, and drug screening [44].

PDTO biobanks have been established for various tumor types, including those from the brain [45, 46], head and neck [47, 48], nasopharynx [49], breast [50], lungs [51, 52], biliary tract [53], liver [4], kidneys [54, 55], esophageal [56], stomach [57, 58], colon [59, 60], rectal [61], pancreas [11, 62], prostate [63, 64], bladder [65], ovarian (Fig. 1.1) [2, 13], neuroendocrine tumors [66], upper tract urothelial carcinoma [67], and colorectal cancer liver metastases [9]. This marks a significant advancement in tumor biology and serves as a crucial resource for research and personalized medicine. Previous studies have examined the recovery rates, survival efficiency, and growth of these organoids following revival in vitro [68]. As a result, PDO biobanks greatly increase the variety of patient samples that can be cultured and analyzed in the laboratory. These biobanks accurately replicate the phenotypic and genetic traits of target organs, providing innovative platforms for investigating cancer development and progression, conducting drug screening, and establishing preclinical models, as previously discussed.

Organoids can better reproduce tumor heterogeneity and drug response. The most comprehensive biobank up to date can be considered as the one established by the Hubrecht Institute, Utrecht University Medical Center, and the Royal Netherlands Academy of Arts and Sciences with more than 1000 organoids from a diversity of organs and diseases such as breast, colon, head and neck tumors, intestinal, liver, lung, ovarian, and pancreatic tumors, along with a large number of intestinal organoids in cystic fibrosis patients [69, 70].

A recent paper explained each and every systemic organoid biobank repository in the world recently in detail [68]. In addition to the ovarian tumor organoid biobank information from Nelson et al. [71] in the reference paper, our group also developed a biobank of ovarian tumor-derived organoids originating from high-grade serous ovarian cancer (HGSOCs) and couple of low-grade serous ovarian cancer (LGSOCs) and clear cell carcinomas. Our biobank covers organoids banked with different germline mutations such as *BRCA1*, *BRCA2*, and *ATM* [2, 13].

1.6 Biomarker Discovery with Patient-Derived Tumor Organoids

Patient-derived organoids offer a more advanced preclinical model than previous systems due to their natural heterogeneity, long-term stability, suitability for high-throughput screening, and improved ability to replicate tumor characteristics [3, 72, 73]. In 2018, it was initially shown that PDOs could potentially forecast treatment responses in cancer patients [22].

Cancer biomarkers play a crucial role in categorizing patients and anticipating their response to treatment. While effective biomarkers have been identified for certain cancers, such as breast cancer, leading to improved drug therapy, many other cancer types still lack reliable predictors of treatment response. Liquid biopsies, especially blood samples, offer a non-invasive method for detecting potential biomarkers, including circulating tumor cells, methylated DNA, cell-free DNA, circulating RNA (such as microRNA), and extracellular vesicles (such as exosomes) [74, 75]. Identifying potential predictive biomarkers will provide insight into therapy sensitivity or resistance, mechanisms of drug resistance, and the effects of combination treatments. The use of organoid models, as demonstrated in various cancer types, will further enhance this understanding. Additionally, the latest revision by the Food and Drug Administration (FDA) endorses the use of alternative in vitro models in cases where animal testing is not necessary, aligning with the principles of the 3Rs approach [76].

Findings from comparisons between ex vivo organoid experiments and clinical trial outcomes suggest that PDOs can reliably mirror patient responses. Therefore, integrating PDOs into personalized medicine approaches holds promise for practical application [77, 78]. Mo et al. found that the in vitro effectiveness of PDOs treated with FOLFOX (folinic acid, fluorouracil, and oxaliplatin) or FOLFIRI (folinic acid, fluorouracil, and irinotecan) is associated with progression-free survival in patients with colorectal liver metastasis (CRLM). This underscores the potential of PDOs to predict chemotherapy responses in CRLM patients within clinical environments [9]. A recent study showed that PDOs effectively assess patient-specific drug responses as ex vivo models in advanced breast cancer, with patients receiving treatments identified as sensitive by PDOs exhibiting positive clinical outcomes [79]. Likewise, a study by Chen and colleagues developed a whole-tumor cell (WTC) ex vivo model for breast cancer to conduct drug profiling across various therapies. Additionally, they validated their model by replicating the treatment regimens of 15 breast cancer patients and discovered that the ex vivo WTC model accurately predicted clinical treatment responses [80]. Hu et al. emphasize the use of PDOs as a model for studying human papillomavirus-related precancerous cervical lesions and cervical cancer. Utilizing these PDOs could help patients with drug resistance identify more effective chemotherapy options, ultimately improving their therapeutic outcomes [81].

1.7 Conclusions and Future Directions

PDTOs represent powerful preclinical models for predicting in vivo drug responses, as they effectively recapitulate both inter-patient and intra-tumoral heterogeneity. At the multi-omics level, they closely mirror not only the tumor type but also the specific tumor sample from which they are derived. This high degree of fidelity enables more comprehensive preclinical evaluation of emerging therapies, which is particularly valuable for rare cancer types. As such, PDTOs hold significant promise as platforms for drug development and retrospective analyses of clinical trials. However, several challenges remain in their application to precision oncology. These include low success rates in establishing PDTOs from certain tumor types, the time-intensive process of generating sufficient quantities for high-throughput drug screening, and the requirement for specialized technical expertise. Advancements in assay miniaturization and automation could substantially improve the scalability and clinical utility of PDTO-based approaches in personalized cancer therapy.

References

1. Driehuis E, Kretzschmar K, Clevers H. Establishment of patient-derived cancer organoids for drug-screening applications. Nat Protoc. 2020;15(10):3380–409.
2. da Costa AA, Somuncu O, Ravindranathan R, Mukkavalli S, Martignetti DB, Nguyen H, et al. Single-stranded DNA gap accumulation is a functional biomarker for USP1 inhibitor sensitivity. Cancer Res. 2024;84(20):3435–46.
3. van de Wetering M, Francies HE, Francis JM, Bounova G, Iorio F, Pronk A, et al. Prospective derivation of a living organoid biobank of colorectal cancer patients. Cell. 2015;161(4):933–45.
4. Broutier L, Mastrogiovanni G, Verstegen MM, Francies HE, Gavarro LM, Bradshaw CR, et al. Human primary liver cancer-derived organoid cultures for disease modeling and drug screening. Nat Med. 2017;23(12):1424–35.
5. Lee SH, Hu W, Matulay JT, Silva MV, Owczarek TB, Kim K, et al. Tumor evolution and drug response in patient-derived organoid models of bladder cancer. Cell. 2018;173(2):515–28 e17.
6. Veninga V, Voest EE. Tumor organoids: opportunities and challenges to guide precision medicine. Cancer Cell. 2021;39(9):1190–201.
7. Thorel L, Perreard M, Florent R, Divoux J, Coffy S, Vincent A, et al. Patient-derived tumor organoids: a new avenue for preclinical research and precision medicine in oncology. Exp Mol Med. 2024;56(7):1531–51.
8. Sato T, Vries RG, Snippert HJ, van de Wetering M, Barker N, Stange DE, et al. Single Lgr5 stem cells build crypt-villus structures in vitro without a mesenchymal niche. Nature. 2009;459(7244):262–5.
9. Mo S, Tang P, Luo W, Zhang L, Li Y, Hu X, et al. Patient-derived organoids from colorectal cancer with paired liver metastasis reveal tumor heterogeneity and predict response to chemotherapy. Adv Sci (Weinh). 2022;9(31):e2204097.
10. Wang HM, Zhang CY, Peng KC, Chen ZX, Su JW, Li YF, et al. Using patient-derived organoids to predict locally advanced or metastatic lung cancer tumor response: a real-world study. Cell Rep Med. 2023;4(2):100911.
11. Tiriac H, Belleau P, Engle DD, Plenker D, Deschenes A, Somerville TDD, et al. Organoid profiling identifies common responders to chemotherapy in pancreatic cancer. Cancer Discov. 2018;8(9):1112–29.
12. Chang KC, Silvestri F, Oliphant MUJ, Martinez-Gakidis MA, Orgill DP, Garber JE, et al. Breast organoid suspension cultures maintain long-term estrogen receptor expression and responsiveness. NPJ Breast Cancer. 2024;10(1):107.
13. Ravindranathan R, Somuncu O, da Costa A, Mukkavalli S, Lamarre BP, Nguyen H, et al. PARG inhibitor sensitivity correlates with accumulation of single-stranded DNA gaps in preclinical models of ovarian cancer. Proc Natl Acad Sci USA. 2024;121(47):e2413954121.
14. Somuncu O, Ashton N, da Costa AA, Mukkavalli S, Ravindranathan R, Lamarre B, et al. Identification of predictive biomarkers of USP1 inhibitor sensitivity using ovarian cancer organoid models. Cancer Res. 2025;85(8_Supplement_1):4004.
15. Brennen WN, Le Magnen C, Karkampouna S, Anselmino N, Bock N, Choo N, et al. Defining the challenges and opportunities for using patient-derived models in prostate cancer research. Prostate. 2024;84(7):623–35.
16. Patel S, Alam A, Pant R, Chattopadhyay S. Wnt signaling and its significance within the tumor microenvironment: novel therapeutic insights. Front Immunol. 2019;10:2872.
17. Xu H, Lyu X, Yi M, Zhao W, Song Y, Wu K. Organoid technology and applications in cancer research. J Hematol Oncol. 2018;11(1):116.
18. Bae J, Choi YS, Cho G, Jang SJ. The patient-derived cancer organoids: promises and challenges as platforms for cancer discovery. Cancers (Basel). 2022;14(9):2144.
19. Taurin S, Alzahrani R, Aloraibi S, Ashi L, Alharmi R, Hassani N. Patient-derived tumor organoids: a preclinical platform for personalized cancer therapy. Transl Oncol. 2025;51:102226.
20. Tong L, Cui W, Zhang B, Fonseca P, Zhao Q, Zhang P, et al. Patient-derived organoids in precision cancer medicine. Med. 2024;5(11):1351–77.

21. Phan N, Hong JJ, Tofig B, Mapua M, Elashoff D, Moatamed NA, et al. A simple high-throughput approach identifies actionable drug sensitivities in patient-derived tumor organoids. Commun Biol. 2019;2:78.
22. Vlachogiannis G, Hedayat S, Vatsiou A, Jamin Y, Fernandez-Mateos J, Khan K, et al. Patient-derived organoids model treatment response of metastatic gastrointestinal cancers. Science. 2018;359(6378):920–6.
23. Gorski JW, Zhang Z, McCorkle JR, DeJohn JM, Wang C, Miller RW, et al. Utilizing patient-derived epithelial ovarian cancer tumor organoids to predict carboplatin resistance. Biomedicine. 2021;9(8):1021.
24. Liu HD, Xia BR, Jin MZ, Lou G. Organoid of ovarian cancer: genomic analysis and drug screening. Clin Transl Oncol. 2020;22(8):1240–51.
25. Schuster B, Junkin M, Kashaf SS, Romero-Calvo I, Kirby K, Matthews J, et al. Automated microfluidic platform for dynamic and combinatorial drug screening of tumor organoids. Nat Commun. 2020;11(1):5271.
26. Nakabori T, Abe Y, Higashi S, Mukai K, Yoshioka R, Morimoto Y, et al. Usefulness of on-site cytology of liver tumor biopsy in specimen sampling for cancer genomic profiling test. Cancer Med. 2023;12(7):7888–92.
27. Pasch CA, Favreau PF, Yueh AE, Babiarz CP, Gillette AA, Sharick JT, et al. Patient-derived cancer organoid cultures to predict sensitivity to chemotherapy and radiation. Clin Cancer Res. 2019;25(17):5376–87.
28. Yoon H, Lee S. Integration of genomic profiling and organoid development in precision oncology. Int J Mol Sci. 2021;23(1):216.
29. Bergdorf KN, Phifer CJ, Bechard ME, Lee MA, McDonald OG, Lee E, et al. Immunofluorescent staining of cancer spheroids and fine-needle aspiration-derived organoids. STAR Protoc. 2021;2(2):100578.
30. Regan JL. Immunofluorescence staining of colorectal cancer patient-derived organoids. Methods Cell Biol. 2022;171:163–71.
31. Besikcioglu HE, Yurteri U, Ye L, Zhang F, Moretti A, Gurcinar IH, et al. Protocol for whole-mount immunofluorescence staining of ECM gel-embedded innervated pancreatic organoids. STAR Protoc. 2024;5(2):103132.
32. McCray T, Richards Z, Marsili J, Prins GS, Nonn L. Handling and assessment of human primary prostate organoid culture. J Vis Exp. 2019;(143). https://doi.org/10.3791/59051.
33. Hill SJ, Decker B, Roberts EA, Horowitz NS, Muto MG, Worley MJ Jr, et al. Prediction of DNA repair inhibitor response in short-term patient-derived ovarian cancer organoids. Cancer Discov. 2018;8(11):1404–21.
34. Bussooa A, Tubbs E, Revol-Cavalier F, Chmayssem A, Alessio M, Cosnier M-L, et al. Real-time monitoring of oxygen levels within thermoplastic organ-on-Chip devices. Biosens Bioelectron X. 2022;11:100198.
35. Dornhof J, Kieninger J, Muralidharan H, Maurer J, Urban GA, Weltin A. Microfluidic organ-on-chip system for multi-analyte monitoring of metabolites in 3D cell cultures. Lab Chip. 2022;22(2):225–39.
36. Braun LM, Lagies S, Klar RFU, Hussung S, Fritsch RM, Kammerer B, et al. Metabolic profiling of early and late recurrent pancreatic ductal adenocarcinoma using patient-derived organoid cultures. Cancers (Basel). 2020;12(6):1440.
37. Walsh AJ, Castellanos JA, Nagathihalli NS, Merchant NB, Skala MC. Optical imaging of drug-induced metabolism changes in murine and human pancreatic cancer organoids reveals heterogeneous drug response. Pancreas. 2016;45(6):863–9.
38. Walsh AJ, Cook RS, Sanders ME, Aurisicchio L, Ciliberto G, Arteaga CL, et al. Quantitative optical imaging of primary tumor organoid metabolism predicts drug response in breast cancer. Cancer Res. 2014;74(18):5184–94.
39. Walsh AJ, Cook RS, Sanders ME, Arteaga CL, Skala MC. Drug response in organoids generated from frozen primary tumor tissues. Sci Rep. 2016;6:18889.

40. Vander Linden C, Corbet C. Reconciling environment-mediated metabolic heterogeneity with the oncogene-driven cancer paradigm in precision oncology. Semin Cell Dev Biol. 2020;98:202–10.
41. Weygand J, Carter S, Salzillo T, Moussalli M, Dai B. Can an organoid recapitulate the metabolome of its parent tissue. A pilot NMR spectroscopy study. J Cancer Prev Curr Res. 2017;7:00307.
42. Yoshizaki H, Ogiso H, Okazaki T, Kiyokawa E. Comparative lipid analysis in the normal and cancerous organoids of MDCK cells. J Biochem. 2016;159(6):573–84.
43. Lindeboom RG, van Voorthuijsen L, Oost KC, Rodríguez-Colman MJ, Luna-Velez MV, Furlan C, et al. Integrative multi-omics analysis of intestinal organoid differentiation. Mol Syst Biol. 2018;14(6):e8227.
44. Li H, Liu H, Chen K. Living biobank-based cancer organoids: prospects and challenges in cancer research. Cancer Biol Med. 2022;19(7):965–82.
45. Jacob F, Salinas RD, Zhang DY, Nguyen PTT, Schnoll JG, Wong SZH, et al. A patient-derived glioblastoma organoid model and biobank recapitulates inter- and intra-tumoral heterogeneity. Cell. 2020;180(1):188–204 e22.
46. Hubert CG, Rivera M, Spangler LC, Wu Q, Mack SC, Prager BC, et al. A three-dimensional organoid culture system derived from human glioblastomas recapitulates the hypoxic gradients and cancer stem cell heterogeneity of tumors found in vivo. Cancer Res. 2016;76(8):2465–77.
47. Millen R, De Kort WWB, Koomen M, van Son GJF, Gobits R, Penning de Vries B, et al. Patient-derived head and neck cancer organoids allow treatment stratification and serve as a tool for biomarker validation and identification. Med. 2023;4(5):290–310.e12.
48. Driehuis E, Kolders S, Spelier S, Lohmussaar K, Willems SM, Devriese LA, et al. Oral mucosal organoids as a potential platform for personalized cancer therapy. Cancer Discov. 2019;9(7):852–71.
49. Wang XW, Xia TL, Tang HC, Liu X, Han R, Zou X, et al. Establishment of a patient-derived organoid model and living biobank for nasopharyngeal carcinoma. Ann Transl Med. 2022;10(9):526.
50. Sachs N, de Ligt J, Kopper O, Gogola E, Bounova G, Weeber F, et al. A living biobank of breast cancer organoids captures disease heterogeneity. Cell. 2018;172(1-2):373–86.e10.
51. Sachs N, Papaspyropoulos A, Zomer-van Ommen DD, Heo I, Bottinger L, Klay D, et al. Long-term expanding human airway organoids for disease modeling. EMBO J. 2019;38(4):e100300.
52. Ebisudani T, Hamamoto J, Togasaki K, Mitsuishi A, Sugihara K, Shinozaki T, et al. Genotype-phenotype mapping of a patient-derived lung cancer organoid biobank identifies NKX2-1-defined Wnt dependency in lung adenocarcinoma. Cell Rep. 2023;42(3):112212.
53. Nasar N, McIntyre SMH, Kalvin HL, Gonen M, Lecomte N, Karnoub E, et al. Patient-derived biliary tract cancer organoid biobank: is personalized medicine utilizing organoids feasible for patients with cholangiocarcinoma? American Society of Clinical Oncology; 2024.
54. Calandrini C, Schutgens F, Oka R, Margaritis T, Candelli T, Mathijsen L, et al. An organoid biobank for childhood kidney cancers that captures disease and tissue heterogeneity. Nat Commun. 2020;11(1):1310.
55. Grassi L, Alfonsi R, Francescangeli F, Signore M, De Angelis ML, Addario A, et al. Organoids as a new model for improving regenerative medicine and cancer personalized therapy in renal diseases. Cell Death Dis. 2019;10(3):201.
56. Karakasheva TA, Gabre JT, Sachdeva UM, Cruz-Acuna R, Lin EW, DeMarshall M, et al. Patient-derived organoids as a platform for modeling a patient's response to chemoradiotherapy in esophageal cancer. Sci Rep. 2021;11(1):21304.
57. Yan HHN, Siu HC, Law S, Ho SL, Yue SSK, Tsui WY, et al. A comprehensive human gastric cancer organoid biobank captures tumor subtype heterogeneity and enables therapeutic screening. Cell Stem Cell. 2018;23(6):882–97 e11.
58. Nanki K, Toshimitsu K, Takano A, Fujii M, Shimokawa M, Ohta Y, et al. Divergent routes toward Wnt and R-spondin niche independency during human gastric carcinogenesis. Cell. 2018;174(4):856–69 e17.

59. Fujii M, Shimokawa M, Date S, Takano A, Matano M, Nanki K, et al. A colorectal tumor organoid library demonstrates progressive loss of niche factor requirements during tumorigenesis. Cell Stem Cell. 2016;18(6):827–38.
60. Smabers LP, Wensink E, Verissimo CS, Koedoot E, Pitsa KC, Huismans MA, et al. Organoids as a biomarker for personalized treatment in metastatic colorectal cancer: drug screen optimization and correlation with patient response. J Exp Clin Cancer Res. 2024;43(1):61.
61. Andel D, Viergever BJ, Peters NA, Elisabeth Raats DA, Schenning-van Schelven SJ, Willem Intven MP, et al. Pre-existing subclones determine radioresistance in rectal cancer organoids. Cell Rep. 2024;43(2):113735.
62. Seino T, Kawasaki S, Shimokawa M, Tamagawa H, Toshimitsu K, Fujii M, et al. Human pancreatic tumor organoids reveal loss of stem cell niche factor dependence during disease progression. Cell Stem Cell. 2018;22(3):454–67.e6.
63. Beshiri ML, Tice CM, Tran C, Nguyen HM, Sowalsky AG, Agarwal S, et al. A PDX/organoid biobank of advanced prostate cancers captures genomic and phenotypic heterogeneity for disease modeling and therapeutic screening. Clin Cancer Res. 2018;24(17):4332–45.
64. Gao D, Vela I, Sboner A, Iaquinta PJ, Karthaus WR, Gopalan A, et al. Organoid cultures derived from patients with advanced prostate cancer. Cell. 2014;159(1):176–87.
65. Radic M, Egger M, Kruithof-de Julio M, Seiler R. Patient-derived organoids in bladder cancer: opportunities and challenges. Eur Urol Focus. 2025;11(1):62–70.
66. Kawasaki K, Toshimitsu K, Matano M, Fujita M, Fujii M, Togasaki K, et al. An organoid biobank of neuroendocrine neoplasms enables genotype-phenotype mapping. Cell. 2020;183(5):1420–35.e21.
67. Li Z, Xu H, Gong Y, Chen W, Zhan Y, Yu L, et al. Patient-derived upper tract urothelial carcinoma organoids as a platform for drug screening. Adv Sci (Weinh). 2022;9(4):e2103999.
68. Xie X, Li X, Song W. Tumor organoid biobank-new platform for medical research. Sci Rep. 2023;13(1):1819.
69. den Hertog J, de Laat SW. Hubrecht institute centennial – from embryos to stem cells. Dev Biol. 2017;428(2):261–3.
70. Rabouille C, Deschamps J. On the shoulders of Hubrecht: from embryos to stem cells. Dev Biol. 2017;428(2):264–72.
71. Nelson L, Tighe A, Golder A, Littler S, Bakker B, Moralli D, et al. A living biobank of ovarian cancer ex vivo models reveals profound mitotic heterogeneity. Nat Commun. 2020;11(1):822.
72. Sato T, Stange DE, Ferrante M, Vries RG, Van Es JH, Van den Brink S, et al. Long-term expansion of epithelial organoids from human colon, adenoma, adenocarcinoma, and Barrett's epithelium. Gastroenterology. 2011;141(5):1762–72.
73. Pampaloni F, Reynaud EG, Stelzer EH. The third dimension bridges the gap between cell culture and live tissue. Nat Rev Mol Cell Biol. 2007;8(10):839–45.
74. Yang C, Xia BR, Jin WL, Lou G. Circulating tumor cells in precision oncology: clinical applications in liquid biopsy and 3D organoid model. Cancer Cell Int. 2019;19:341.
75. Iovanna J. Implementing biological markers as a tool to guide clinical care of patients with pancreatic cancer. Transl Oncol. 2021;14(1):100965.
76. Han JJ. FDA modernization act 2.0 allows for alternatives to animal testing. Artif Organs. 2023;47(3):449–50.
77. Love JR, Karthaus WR. Next-generation modeling of cancer using organoids. Cold Spring Harb Perspect Med. 2024;14(6):a041380.
78. LeSavage BL, Suhar RA, Broguiere N, Lutolf MP, Heilshorn SC. Next-generation cancer organoids. Nat Mater. 2022;21(2):143–59.
79. Chen P, Zhang X, Ding R, Yang L, Lyu X, Zeng J, et al. Patient-derived organoids can guide personalized-therapies for patients with advanced breast cancer. Adv Sci (Weinh). 2021;8(22):e2101176.

80. Chen X, Sifakis EG, Robertson S, Neo SY, Jun SH, Tong L, et al. Breast cancer patient-derived whole-tumor cell culture model for efficient drug profiling and treatment response prediction. Proc Natl Acad Sci USA. 2023;120(1):e2209856120.
81. Hu B, Wang R, Wu D, Long R, Fan J, Hu Z, et al. A promising new model: establishment of patient-derived organoid models covering HPV-related cervical pre-cancerous lesions and their cancers. Adv Sci (Weinh). 2024;11(12):e2302340.

Chapter 2
Bladder Organoids for Healthy and Diseased Tissue Modeling

Selim Sevim and Dilara Akbulut

Abstract Bladder organoids have emerged as promising three-dimensional (3D) in vitro models that faithfully recapitulate the structural and functional characteristics of human bladder tissue. Derived from primary urothelial cells, stem cells, or patient tumor samples, these organoids provide a physiologically relevant platform for studying both healthy bladder biology and disease progression, including cancer, infection, and inflammatory conditions. By maintaining key features such as cellular heterogeneity, tissue polarity, and urothelial differentiation, bladder organoids enable the investigation of normal tissue regeneration, response to injury, and molecular mechanisms underlying pathological states. Moreover, disease-specific bladder organoids, particularly those derived from patients with bladder cancer, offer valuable tools for drug screening, biomarker discovery, and precision medicine applications. This review highlights recent advances in bladder organoid culture systems, their applications in modeling bladder disorders, and ongoing efforts to integrate them into translational research and personalized therapeutic strategies.

Keywords Bladder organoids · Personalized medicine · Bladder cancer

2.1 Introduction

Bladder cancer comprises the sixth most common cancer type in the USA, with an estimate of more than 80,000 new diagnosed cases every year [1].

S. Sevim
Cancer Early Detection Advanced Research Center, Knight Cancer Institute, Oregon Health and Science University, Portland, OR, USA

D. Akbulut (✉)
Department of Pathology, Memorial Sloan Kettering Cancer Center, New York, NY, USA
e-mail: akbulutd@mskcc.org

O. S. Somuncu (ed.), *Organoids for Preclinical Disease Modeling and Personalized Medicine*, Stem Cell Biology and Regenerative Medicine 78,
https://doi.org/10.1007/978-3-032-10024-5_2

For staging and treatment purposes, bladder cancer is primarily categorized into muscle-invasive and non-muscle-invasive types. Approximately, three-quarters of non-muscle-invasive bladder cancer cases are treated with a combination of transurethral resection and intravesical chemotherapy or intravesical BCG [2]. However, a considerable proportion of these cases either recur or progress to muscle-invasive bladder cancer [3, 4]. The remaining approximately one-quarter of patients with muscle-invasive disease are treated with radical cystectomy following neoadjuvant chemotherapy in some cases, although these patients often experience high mortality and morbidity rates [5, 6]. The marked intra-tumoral heterogeneity of bladder cancer, as demonstrated by molecular classifications, complicates treatment by contributing to tumor recurrence and progression [7].

2.1.1 Bladder Cancer Subtypes

Urothelial carcinoma is the most common histologic type of bladder cancer, followed by pure squamous cell carcinoma (<5%) and adenocarcinoma (<1%). About 25% of invasive urothelial carcinoma cases show different histopathologic morphologies along with the conventional urothelial morphology. Different epithelial lineages could be represented as divergent differentiation, such as squamous, glandular, trophoblastic, neuroendocrine, or Mullerian differentiation. Various histologic "subtypes" are important to recognize as they have different molecular drivers and affect the prognosis, treatment selection, and response. Morphologic subtypes include micropapillary, nested, tubular/microcystic, plasmacytoid, sarcomatoid, lipid-rich, clear cell, lymphoepithelioma-like, giant cell, and poorly differentiated urothelial carcinoma [8].

2.1.2 Mutational Profile

Lining urothelium harbors a rich mutational profile which creates a field effect [9]. Comprehensive molecular analyses of bladder cancer demonstrated focal deletions of tumor suppressor genes (*CDKN2A, RB1, NCOR1, PTEN*) as well as amplification of oncogenes (*E2F3, PPARG, CCND1, CCNE1, MYC, and MDM2*) [10]. APOBEC mutation signature, with high somatic mutation rate, was identified in most bladder carcinoma cases in the TCGA cohort [10]. *TERT* promoter mutation is the most common alteration in urothelial carcinoma that is believed to occur early in the disease evolution and is encountered independent from the disease stage, or the histomorphological subtype [11, 12]. Somatic recurrent mutations identified in urothelial carcinoma play role in cell cycle regulation (*TP53, RB1, CDKN2A, CDKN1A*), chromatin modification (*ARID1A, KDM6A, KMT2D, KMT2C, KMT2A, EP300, CREBBP*), receptor tyrosine kinase/RAS signaling (*FGFR3, ERBB2, ERBB3, KRAS, HRAS*), the PI3K/AKT pathway (*PIK3CA, TSC1*), and other

Table 2.1 Mutational profile of bladder cancer [15]

Gene	Prevalence (%)
TP53	49
MLL2	27
ARID1A	25
KDM6A	24
PIK3CA	20
EP300	15
CDKN1A	14
RB1	13
ERCC2	12
FGFR3	12
STAG2	11
ERBB3	11
FBXW7	10
RXRA	9
ELF3	8
NFE2L2	8
TSC1	8
KLF5	8
TXNIP	7
FOXQ1	5
CDKN2A	5
RHOB	5
FOXA1	5
PAIP1	5
BTG2	5
HRAS	5
ZFP36L1	5
RHOA	4
CCND3	4

functions (*STAG2, ATM, FAT1, FBXW7*) [13, 14]. *FGFR3* mutations and oncogenic fusions are frequently identified in bladder cancer, especially in the papillary lesions [12] (Table 2.1).

2.1.3 Molecular and Morphologic Diversity

Bladder carcinoma is unique with diverse morphologic variation supported by distinct molecular alterations, with potential diagnostic, therapeutic, and prognostic implications [16]. For example, micropapillary subtype of urothelial carcinomas are enriched with ERBB2 gene alterations, show a worse prognosis compared to traditional urothelial carcinoma cases, and could be targeted with anti-HER2 agents [17]. Plasmacytoid subtype harbors the pathognomonic CDH1 truncating mutation and has a higher cancer-specific mortality rate [18]. FDA-approved anti-FGFR agents have shown efficacy in advanced disease with FGFR2 or FGFR3 mutation/translocation identified patients [19]. Identification of distinct molecular alterations will further advance the understanding of the evolution of disease and potentially identify targetable mutations, and biomarkers for early detection/screening purposes.

2.2 Organoids in Bladder Cancer

The story and foundations of organoids date back approximately 120 years ago [20]. In the referred study, Wilson and colleagues demonstrated that sponge cells could form organized structures with regenerative potential. In 1961, the term "pluripotent stem cell" was introduced [21], and in the late 1990s and early 2000s, the use of human embryonic stem cells and induced pluripotent stem cells changed the perspective on the fate of these cell groups [22, 23]. The use of intestinal organoids expressing LGR5 receptor for the first time in 2009 accelerated the development of this technology to the present day [24]. Organoid structures can provide critical insights by reflecting the characteristics of stem cells and the organs they form, enabling the proliferation of these cells, mimicking cellular distribution, and representing the regeneration processes of corresponding organs [25]. In summary, organoids are miniature 3D structures that can be produced from human embryonic stem cells, induced pluripotent stem cells, and adult stem cells [26], and they serve as tools for disease research in the hands of scientists. These special structure compounds replicate features such as cell-cell interaction, cell-matrix interaction, and genetic profiles, thus can be used as models to predict disease progression [27–29].

To improve therapeutic outcomes, treatment approaches must be individualized, identifying the most effective compound structure as a cornerstone for each patient. Recent advancements in medicine have shifted toward a more personalized approach, particularly with precision medicine. This approach tailors treatments based on the unique genetic, environmental, and lifestyle factors of each patient, ensuring more targeted and effective therapies. Drug responses observed in organoid cultures derived from bladder cancer cells can be validated through organoid-derived xenograft models, which can then be applied to optimize treatment. In line with these advancements, organoid technology primarily addresses four key areas: disease modeling, drug studies, precision medicine, and regenerative medicine [28, 30]. Additionally, organoid studies are instrumental in identifying genetic mutations, discovering novel biomarkers, and unraveling the molecular mechanisms specific to bladder cancer. By providing a more accurate and personalized platform for studying bladder cancer, organoids enable the exploration of therapeutic targets and allow for a deeper understanding of tumor behavior and progression of this heterogeneous disease. This approach not only accelerates the development of more

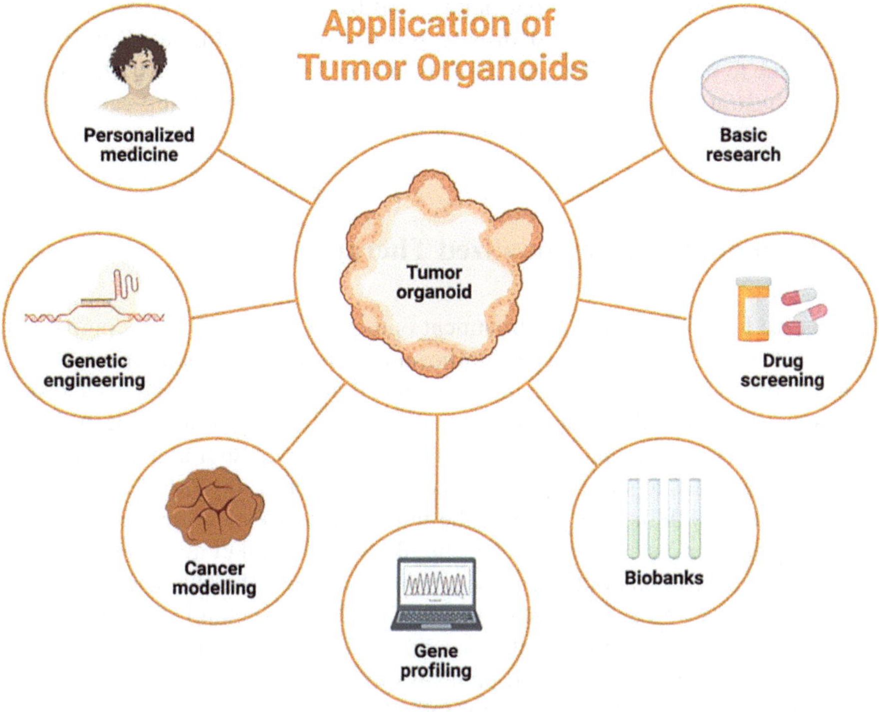

Fig. 2.1 Application of tumor organoids. (Created with BioRender.com)

effective treatments but also facilitates the creation of personalized therapies that are tailored to the genetic and molecular profiles of individual bladder cancer patients, ultimately leading to more targeted and successful treatment options [28, 30, 31]. Main usage areas of organoids were highlighted in Fig. 2.1.

Bladder cancer-based organoids are dynamic structures capable of generating valuable insights into various areas, including cell-cell interactions, tumor characteristics, and drug-related therapeutic investigations. In their study, Medle et al. have started with cell-lines and typically included both high-grade and low-grade morphologies, as well as examples of muscle-invasive and non-muscle-invasive tumors, with a reported successful establishment rate of 33–100% [32].

Results in the literature demonstrated that bladder organoids and organoid-derived xenografts can replicate a wide range of histopathological and molecular features of the original tumor, reflecting the heterogeneity of bladder tumors [31, 33–35]. Common genetic alterations seen in bladder tumors, such as ARID1A, KMT2C, KMT2D, KDM6A, FGFR3, ERBB2, and TP53, were identified in the study of Lee et al., with the help of a large cohort of patient-derived organoids [33]. Clonal relationship of organoids with original tumor was supported using linear and branched phylogenetic trees. Interestingly, the ability of most organoid lines to transition between basal and luminal states based on marker phenotype or RNA expression was observed. This plasticity could be an inherent dynamic nature of organoids, reflecting the natural disease process of luminal/basal switching, or could be due to the lack of stromal elements in vitro [33]. Yoshida et al., for the first time, demonstrated proliferation of bladder cancer cells via the Wnt/β-catenin pathway activation in the organoid cultures [36].

2.3 Organoids for Personalized Therapy

The relative lack of personalized treatment options in bladder cancer, compared to other solid tumor types, presents a significant research opportunity, especially with the use of organoids. Studies with dose-dependent drug response analyses showed a correlation between the response and the mutational profile. Heterogeneous response results with standard of care drugs were reported, highlighting the potential use of organoids to stratify patients according to sensitivity [31, 33–35].

In the study by Minoli et al. [31], the importance of patient-derived organoids [37–40] in bladder cancer research was highlighted. Minoli et al. [31] emphasized that patient-derived organoids have the potential to reflect the characteristics of the parental tumor as well as in the other studies [10, 36]. This claim was supported by two longitudinal retrospective analyses conducted within the study. In these analyses, drugs targeting FDA-approved pathways frequently altered in bladder cancer, such as FGFR, EGFR/ERBB2, and mTOR, were utilized [41]. Minoli et al. concluded that patient-derived bladder cancer organoids can effectively represent tumor heterogeneity, reflect the mutational burden, and serve as valuable tools in drug sensitivity and resistance screenings [31].

In the study by Burgués et al., multi-agent assays were tested on patient-derived organoid models with promising results [42]. Jiang et al. suggested potential use of chimeric antigen receptor-T (CAR-T) cell therapies with specific antigen recognition by CAR-T cells when they are co-cultured with patient-derived organoids [34]. In addition to that, the study team concluded that bladder cancer organoids may guide the therapy choice according to the sensitivity status and immune-related responses in context of the usage of poly adenosine diphosphate-ribose polymerase inhibitor (PARPi) drugs [34]. As a result, the beneficial use of organoids has been demonstrated in numerous independent studies across various areas of personalized medicine [43–53].

Patient-derived organoids also enable the creation of extensive biobanks [35, 54–57]. Data obtained from biobanks have identified potential new therapeutic options, which could contribute as an effective complementation in the treatment of bladder cancer [58] or may guide scientists to a better understanding of disease evolution.

2.4 AI-Enhanced Organoid Models

With the advancement of artificial intelligence (AI) technologies, significant progress has also been made in organoid technology [59, 60]. The ability of organoids to facilitate rapid drug testing, when combined with AI technologies, microfluidics, and image analysis, has enabled the rapid evaluation of drugs and opened pathways for personalized treatments [61, 62]. In Kong et al.'s study, a network was developed with the aim of considering protein-protein interactions [63]. In this work, organoids from 77 bladder cancer patients treated with cisplatin were utilized, alongside with colorectal cancer models. This study, which integrated transcriptomic data, grouped individuals as "responder" and "non-responder," based on the efficacy status of cisplatin chemotherapy. Considering that fewer than half of bladder cancer patients respond to cisplatin [64], the organoid-based network was highlighted as a potential tool for enabling personalized therapies and accelerating the discovery of cancer biomarkers in bladder cancer treatment.

2.5 Conclusion

In conclusion, organoids, especially patient-derived tumor organoids, have transformed research in areas like cancer biology, regenerative medicine, and precision oncology. Derived from normal bladder tissue or urothelial carcinoma, these structures offer a more efficient alternative to traditional cell lines and enable prolonged culture. In clinical point of view, organoids could have a potential to help tailor treatments for metastatic bladder cancer or high-risk non-muscle-invasive bladder cancer patients. Additionally, they provide valuable insights into tumor evolution and heterogeneity, with potential to improve survival outcomes and guide drug usage strategies including newly discovered ones. Lastly, the ethical implications of organoids should not be overlooked.

Possible disadvantages of using organoid structures for drug screening are the challenge to reproduce exact microenvironment, effect of used culture mediums, tumor heterogeneity, and contamination. Another setback is the lack of capacity to test BCG-like agents as those require the immunologic system to act. Lastly, donor awareness regarding experimental studies and ethical considerations in organoid commercialization must be addressed [65].

References

1. Siegel RL, Miller KD, Jemal A. Cancer statistics, 2020. CA Cancer J Clin. 2020;70(1):7–30.
2. Babjuk M, Burger M, Comperat EM, Gontero P, Mostafid AH, Palou J, et al. European Association of Urology guidelines on non-muscle-invasive bladder cancer (TaT1 and carcinoma in situ) – 2019 update. Eur Urol. 2019;76(5):639–57.

3. Soukup V, Capoun O, Cohen D, Hernandez V, Babjuk M, Burger M, et al. Prognostic performance and reproducibility of the 1973 and 2004/2016 World Health Organization grading classification systems in non-muscle-invasive bladder cancer: a European Association of Urology non-muscle invasive bladder cancer guidelines panel systematic review. Eur Urol. 2017;72(5):801–13.
4. Matulewicz RS, Steinberg GD. Non-muscle-invasive bladder cancer: overview and contemporary treatment landscape of neoadjuvant chemoablative therapies. Rev Urol. 2020;22(2):43–51.
5. Facchini G, Cavaliere C, Romis L, Mordente S, Facchini S, Iovane G, et al. Advanced/metastatic bladder cancer: current status and future directions. Eur Rev Med Pharmacol Sci. 2020;24(22):11536–52.
6. Witjes JA, Bruins HM, Cathomas R, Comperat EM, Cowan NC, Gakis G, et al. European Association of Urology guidelines on muscle-invasive and metastatic bladder cancer: summary of the 2020 guidelines. Eur Urol. 2021;79(1):82–104.
7. Kamoun A, de Reyniès A, Allory Y, Sjödahl G, Robertson AG, Seiler R, et al. A consensus molecular classification of muscle-invasive bladder cancer. Eur Urol. 2020;77(4):420–33.
8. WHO Classification of Tumours Editorial Board. Urinary and male genital tumours. Lyon: International Agency for Research on Cancer; 2022.
9. Faltas BM, Prandi D, Tagawa ST, Molina AM, Nanus DM, Sternberg C, et al. Clonal evolution of chemotherapy-resistant urothelial carcinoma. Nat Genet. 2016;48(12):1490–9.
10. Robertson AG, Kim J, Al-Ahmadie H, Bellmunt J, Guo G, Cherniack AD, et al. Comprehensive molecular characterization of muscle-invasive bladder cancer. Cell. 2017;171(3):540–56.e25.
11. Kinde I, Munari E, Faraj SF, Hruban RH, Schoenberg M, Bivalacqua T, et al. TERT promoter mutations occur early in urothelial neoplasia and are biomarkers of early disease and disease recurrence in urine. Cancer Res. 2013;73(24):7162–7.
12. Pietzak EJ, Bagrodia A, Cha EK, Drill EN, Iyer G, Isharwal S, et al. Next-generation sequencing of nonmuscle invasive bladder cancer reveals potential biomarkers and rational therapeutic targets. Eur Urol. 2017;72(6):952–9.
13. Network CGAR. Comprehensive molecular characterization of urothelial bladder carcinoma. Nature. 2014;507(7492):315–22.
14. Kim PH, Cha EK, Sfakianos JP, Iyer G, Zabor EC, Scott SN, et al. Genomic predictors of survival in patients with high-grade urothelial carcinoma of the bladder. Eur Urol. 2015;67(2):198–201.
15. Cancer Genome Atlas Research Network. Comprehensive molecular characterization of urothelial bladder carcinoma. Nature. 2014;507(7492):315–22.
16. Gandhi J, Chen JF, Al-Ahmadie H. Urothelial carcinoma: divergent differentiation and morphologic subtypes. Surg Pathol Clin. 2022;15(4):641–59.
17. Guo CC, Dadhania V, Zhang L, Majewski T, Bondaruk J, Sykulski M, et al. Gene expression profile of the clinically aggressive micropapillary variant of bladder cancer. Eur Urol. 2016;70(4):611–20.
18. Al-Ahmadie HA, Iyer G, Lee BH, Scott SN, Mehra R, Bagrodia A, et al. Frequent somatic CDH1 loss-of-function mutations in plasmacytoid variant bladder cancer. Nat Genet. 2016;48(4):356–8.
19. Pederzoli F, Bandini M, Marandino L, Ali SM, Madison R, Chung J, et al. Targetable gene fusions and aberrations in genitourinary oncology. Nat Rev Urol. 2020;17(11):613–25.
20. Wilson H. A new method by which sponges may be artificially reared. Science. 1907;25(649):912–5.
21. Till JE, Mc CE. A direct measurement of the radiation sensitivity of normal mouse bone marrow cells. Radiat Res. 1961;14:213–22.
22. Takahashi K, Yamanaka S. Induction of pluripotent stem cells from mouse embryonic and adult fibroblast cultures by defined factors. Cell. 2006;126(4):663–76.
23. Thomson JA, Itskovitz-Eldor J, Shapiro SS, Waknitz MA, Swiergiel JJ, Marshall VS, et al. Embryonic stem cell lines derived from human blastocysts. Science. 1998;282(5391):1145–7.

24. Sato T, Vries RG, Snippert HJ, Van De Wetering M, Barker N, Stange DE, et al. Single Lgr5 stem cells build crypt-villus structures in vitro without a mesenchymal niche. Nature. 2009;459(7244):262–5.
25. Yi SA, Zhang Y, Rathnam C, Pongkulapa T, Lee KB. Bioengineering approaches for the advanced organoid research. Adv Mater. 2021;33(45):e2007949.
26. Garreta E, Kamm RD, de Sousa C, Lopes SM, Lancaster MA, Weiss R, Trepat X, et al. Rethinking organoid technology through bioengineering. Nat Mater. 2021;20(2):145–55.
27. Chen B, Du C, Wang M, Guo J, Liu X. Organoids as preclinical models of human disease: progress and applications. Med Rev. 2024;4(2):129–53.
28. Yao Q, Cheng S, Pan Q, Yu J, Cao G, Li L, et al. Organoids: development and applications in disease models, drug discovery, precision medicine, and regenerative medicine. MedComm (2020). 2024;5(10):e735.
29. Zhang YS, Aleman J, Shin SR, Kilic T, Kim D, Mousavi Shaegh SA, et al. Multisensor-integrated organs-on-chips platform for automated and continual in situ monitoring of organoid behaviors. Proc Natl Acad Sci USA. 2017;114(12):E2293–E302.
30. Rossi G, Manfrin A, Lutolf MP. Progress and potential in organoid research. Nat Rev Genet. 2018;19(11):671–87.
31. Minoli M, Cantore T, Hanhart D, Kiener M, Fedrizzi T, La Manna F, et al. Bladder cancer organoids as a functional system to model different disease stages and therapy response. Nat Commun. 2023;14(1):2214.
32. Medle B, Sjodahl G, Eriksson P, Liedberg F, Hoglund M, Bernardo C. Patient-derived bladder cancer organoid models in tumor biology and drug testing: a systematic review. Cancers (Basel). 2022;14(9):2062.
33. Lee SH, Hu W, Matulay JT, Silva MV, Owczarek TB, Kim K, et al. Tumor evolution and drug response in patient-derived organoid models of bladder cancer. Cell. 2018;173(2):515–28 e17.
34. Jiang Y, Sun X, Song X, Li Z, Zhang P, Zhang W, et al. Patient-derived bladder cancer organoid model to predict sensitivity and feasibility of tailored precision therapy. Curr Urol. 2023;17(4):221–8.
35. Mullenders J, de Jongh E, Brousali A, Roosen M, Blom JPA, Begthel H, et al. Mouse and human urothelial cancer organoids: a tool for bladder cancer research. Proc Natl Acad Sci USA. 2019;116(10):4567–74.
36. Yoshida T, Sopko NA, Kates M, Liu X, Joice G, McConkey DJ, et al. Three-dimensional organoid culture reveals involvement of Wnt/beta-catenin pathway in proliferation of bladder cancer cells. Oncotarget. 2018;9(13):11060–70.
37. Hill SJ, Decker B, Roberts EA, Horowitz NS, Muto MG, Worley MJ Jr, et al. Prediction of DNA repair inhibitor response in short-term patient-derived ovarian cancer organoids. Cancer Discov. 2018;8(11):1404–21.
38. Pasch CA, Favreau PF, Yueh AE, Babiarz CP, Gillette AA, Sharick JT, et al. Patient-derived cancer organoid cultures to predict sensitivity to chemotherapy and radiation. Clin Cancer Res. 2019;25(17):5376–87.
39. Vlachogiannis G, Hedayat S, Vatsiou A, Jamin Y, Fernandez-Mateos J, Khan K, et al. Patient-derived organoids model treatment response of metastatic gastrointestinal cancers. Science. 2018;359(6378):920–6.
40. Yao Y, Xu X, Yang L, Zhu J, Wan J, Shen L, et al. Patient-derived organoids predict chemoradiation responses of locally advanced rectal cancer. Cell Stem Cell. 2020;26(1):17–26.e6.
41. Krentel F, Singer F, Rosano-Gonzalez ML, Gibb EA, Liu Y, Davicioni E, et al. A showcase study on personalized in silico drug response prediction based on the genetic landscape of muscle invasive bladder cancer. Sci Rep. 2021;11(1):5849.
42. Burgués JP, Gómez L, Pontones JL, Vera CD, Jiménez-Cruz JF, Ozonas M. A chemosensitivity test for superficial bladder cancer based on three-dimensional culture of tumour spheroids. Eur Urol. 2007;51(4):962–9; discussion 9–70

43. Xie R, Cheng L, Huang M, Huang L, Chen Z, Zhang Q, et al. NAT10 drives cisplatin chemoresistance by enhancing ac4C-associated DNA repair in bladder cancer. Cancer Res. 2023;83(10):1666–83.
44. Shen L, Zhang J, Zheng Z, Yang F, Liu S, Wu Y, et al. PHGDH inhibits ferroptosis and promotes malignant progression by upregulating SLC7A11 in bladder cancer. Int J Biol Sci. 2022;18(14):5459–74.
45. Wang M, Chen X, Tan P, Wang Y, Pan X, Lin T, et al. Acquired semi-squamatization during chemotherapy suggests differentiation as a therapeutic strategy for bladder cancer. Cancer Cell. 2022;40(9):1044–59.e8.
46. Hong X, Chen X, Wang H, Xu Q, Xiao K, Zhang Y, et al. A HER2-targeted antibody-drug conjugate, RC48-ADC, exerted promising antitumor efficacy and safety with intravesical instillation in preclinical models of bladder cancer. Adv Sci (Weinh). 2023;10(32):e2302377.
47. Kim E, Choi S, Kang B, Kong J, Kim Y, Yoon WH, et al. Creation of bladder assembloids mimicking tissue regeneration and cancer. Nature. 2020;588(7839):664–9.
48. Cai EY, Garcia J, Liu Y, Vakar-Lopez F, Arora S, Nguyen HM, et al. A bladder cancer patient-derived xenograft displays aggressive growth dynamics in vivo and in organoid culture. Sci Rep. 2021;11(1):4609.
49. Lu X, Hu C, Duan L, Chen K, Hao H, He Y. Establishment of matched bladder cancer PDX and PDX-derived organoid model and evaluation of anti-tumor efficacy of abemaciclib. Clin Transl Oncol. 2025;27(5):2207–19.
50. Merrill NM, Kaffenberger SD, Bao L, Vandecan N, Goo L, Apfel A, et al. Integrative drug screening and multiomic characterization of patient-derived bladder cancer organoids reveal novel molecular correlates of gemcitabine response. Eur Urol. 2024;86(5):434–44.
51. Wang N, Chen RX, Deng MH, Wei WS, Zhou ZH, Ning K, et al. m(5)C-dependent cross-regulation between nuclear reader ALYREF and writer NSUN2 promotes urothelial bladder cancer malignancy through facilitating RABL6/TK1 mRNAs splicing and stabilization. Cell Death Dis. 2023;14(2):139.
52. Tucci FA, Pennisi R, Rigiracciolo DC, Filippone MG, Bonfanti R, Romeo F, et al. Loss of NUMB drives aggressive bladder cancer via a RHOA/ROCK/YAP signaling axis. Nat Commun. 2024;15(1):10378.
53. Garioni M, Tschan VJ, Blukacz L, Nuciforo S, Parmentier R, Roma L, et al. Patient-derived organoids identify tailored therapeutic options and determinants of plasticity in sarcomatoid urothelial bladder cancer. NPJ Precis Oncol. 2023;7(1):112.
54. Jacob F, Salinas RD, Zhang DY, Nguyen PTT, Schnoll JG, Wong SZH, et al. A patient-derived glioblastoma organoid model and biobank recapitulates inter- and intra-tumoral heterogeneity. Cell. 2020;180(1):188–204.e22.
55. van de Wetering M, Francies HE, Francis JM, Bounova G, Iorio F, Pronk A, et al. Prospective derivation of a living organoid biobank of colorectal cancer patients. Cell. 2015;161(4):933–45.
56. Farin HF, Mosa MH, Ndreshkjana B, Grebbin BM, Ritter B, Menche C, et al. Colorectal cancer organoid-stroma biobank allows subtype-specific assessment of individualized therapy responses. Cancer Discov. 2023;13(10):2192–211.
57. Sachs N, de Ligt J, Kopper O, Gogola E, Bounova G, Weeber F, et al. A living biobank of breast cancer organoids captures disease heterogeneity. Cell. 2018;172(1–2):373–86.e10.
58. Tan P, Wang M, Zhong A, Wang Y, Du J, Wang J, et al. SRT1720 inhibits the growth of bladder cancer in organoids and murine models through the SIRT1-HIF axis. Oncogene. 2021;40(42):6081–92.
59. Bai L, Wu Y, Li G, Zhang W, Zhang H, Su J. AI-enabled organoids: construction, analysis, and application. Bioact Mater. 2024;31:525–48.
60. Bentwich I. Pharma's Bio-AI revolution. Drug Discov Today. 2023;28(5):103515.
61. Deng S, Li C, Cao J, Cui Z, Du J, Fu Z, et al. Organ-on-a-chip meets artificial intelligence in drug evaluation. Theranostics. 2023;13(13):4526–58.

62. Singaraju JP, Kadiresan A, Bhoi RK, Gomez AH, Ma Z, Yang H. Organalysis: multifunctional image preprocessing and analysis software for cardiac organoid studies. Tissue Eng Part C Methods. 2023;29(12):572–82.
63. Kong J, Lee H, Kim D, Han SK, Ha D, Shin K, et al. Network-based machine learning in colorectal and bladder organoid models predicts anti-cancer drug efficacy in patients. Nat Commun. 2020;11(1):5485.
64. Choi W, Porten S, Kim S, Willis D, Plimack ER, Hoffman-Censits J, et al. Identification of distinct basal and luminal subtypes of muscle-invasive bladder cancer with different sensitivities to frontline chemotherapy. Cancer Cell. 2014;25(2):152–65.
65. de Jongh D, Massey EK, VANGUARD Consortium, Bunnik EM. Organoids: a systematic review of ethical issues. Stem Cell Res Ther. 2022;13(1):337.

Chapter 3
The Role of Organoids for Investigating Immunological Function, Disease, and Therapy

Alexander Cicala

Abstract Organoids lie in the middle ground between the full complexity of in vivo organs and traditional two-dimensional (2D) cell cultures. Due to this, organoid models allow researchers to alter the genetic backgrounds of cells more easily, less expensively, and more quickly than generating new mouse models. The advantages of an organoid model vs. an in vivo model may be more obvious than that of a competing in vitro system, e.g., 2D cell lines, but that belies the crucial differences between the two. For instance, because organoids exist in three-dimensional (3D) hydrogels, mechanosensitive signaling pathways are prevented from being activated (as what occurs when cells are in direct contact with plastic or glass well bottoms) and causing differentiation away from pluripotent progenitors. As a result, organoids preserve more cellular diversity and original tissue function than two-dimensional cell cultures do. Furthermore, transplantation of organoids into in vivo systems yields tumors that develop more stroma and stromal components when compared to transplanted tumors from 2D cell lines. Such tumors, by contrast, tend to grow as dense collections of tumor cells: a growth pattern that is histologically less similar to tumors that originated *de novo*. While deep exploration of these subjects is outside the scope of this chapter, which is dedicated to the advantages of applying organoid models to immunological contexts, the results of these properties of organoids hold meaningful implications for the applications of organoids in immunological study and clinical practice. The combined ability of organoids to be easily manipulated, to recapitulate tissue dynamics, and to generate tumors more faithfully to real disease has made organoids a crucial tool for investigating the immune system's function, role in disease, and therapeutics. This chapter will elucidate research contexts in the aforementioned areas into which organoids have

---

A. Cicala (✉)
Renaissance School of Medicine, Stony Brook University, Stony Brook, NY, USA

Stony Brook Cancer Center, Stony Brook Medicine, Stony Brook University, Stony Brook, NY, USA
e-mail: alexander.cicala@stonybrookmedicine.edu

O. S. Somuncu (ed.), *Organoids for Preclinical Disease Modeling and Personalized Medicine*, Stem Cell Biology and Regenerative Medicine 78,
https://doi.org/10.1007/978-3-032-10024-5_3

carved out an irreplaceable niche with evidence from recent studies that demonstrate their advantages.

Keywords Immune crosstalk · Patient-derived organoids · Personalized vaccines

3.1 Introduction

3.1.1 Recapitulating Tissue-Specific Immune Microenvironments

The first advantage of organoid models is that they better maintain the complexity of the tissue microenvironment compared to traditional two-dimensional cell cultures. Since organoids are composed of the full variety of cell types from their parent tissue, secreted extracellular molecules and cytokines are preserved from in vivo tumor microenvironments. Salahudeen et al. confirmed this when they found that organoid cultures retain diverse tumor microenvironmental components without the need for reconstitution [1]. Additionally, it is known that organoids maintain differential expression of immunomodulatory molecules present in primary tissues [2]. Corroborating this, Neal et al. showed the ability to propagate patient-derived organoids with native embedded immune cells, including T cells, B cells, natural killer (NK) cells, and macrophages, preserving the original tumor T cell receptor spectrum [3]. Given the fickle nature of immune cells, this could not have been accomplished by the organoid culture without proper immunomodulatory cytokine production.

Moreover, the hydrogel itself that organoids are cultured in provides another variable to manipulate. Panciera et al. recently demonstrated this not in the context of immune function, but the study still proves the viability of tuning matrix stiffness to manipulate cell behavior [4]. Given that tumors and other pathological processes alter tissue stiffness [5, 6], using different hydrogel stiffnesses to study the impacts of this physical property of tissues on immune cell function would be an additional benefit of organoid models.

In contrast, conventional cultures typically use established cell lines that lack the three-dimensional organization, immune diversity, and patient specificity observed in organoid systems [7, 8]. Overall, studies describe organoids as providing a more physiologically relevant architecture, with preserved microenvironment complexity and immune cell diversity, thereby offering a platform better suited for investigating immune cell interactions and responses.

3.1.2 Modeling Immune/Parenchymal Crosstalk

More than capturing the immune microenvironment, organoid cultures better model the interactions between immune cells and organoid cells of various identities (e.g., epithelial cells, endothelial cells, fibroblasts, or tumor cells), allowing for better study in detail. Chan and Ewald (2022) developed a 3D organotypic co-culture method that captures interactions between NK cells and tumor organoids [9]. They reported that this approach allows for the modeling of key aspects of metastatic biology and screening for the effectiveness of agents that stimulate NK cell cytotoxicity.

Another critical interaction, and perhaps the most famous one in the current paradigm, between immune cells and tumor cells is *immune checkpoint signaling*. Roper and Shen highlighted that 3D cultures preserve immune cell diversity and allow for the modeling of immune checkpoint blockade [10]. They suggested that this capability could be crucial for studying the efficacy of immunotherapies and understanding complex interactions between tumor cells and various immune cell types. Scognamiglio et al. provide proof-of-principal on this subject, by detailing preserved PD-L1 expression and measurable reductions in organoid diameter—with corresponding increases in cell death—following nivolumab treatment [11]. Zhang et al. demonstrate that organoids yield dose-dependent assessments of T cell activation and tumor cell killing, thereby establishing a closer correspondence to in vivo responses [12]. In contrast, 2D cultures generally show limited retention of immune components, altered checkpoint expression, and less reliable therapeutic readouts [3].

Immune-stroma or immune-tumor interactions are central to many immune processes, including tissue repair [13–15], and immune surveillance [13, 16], or evasion by pathogens and tumors alike [16, 17]. Organoids provide a more dynamic and physiologically relevant setting for studying these interactions compared to 2D models.

3.1.3 Deciphering Disease-Specific Immune Kinetics

Organoids derived from stem cells or progenitor cells can be used to study the differentiation of immune cells within specific tissues, providing insights into the early development of the immune system. For example, Jowett et al. demonstrated that both murine and human organoid models could capture tissue-specific maturation of innate lymphoid cell precursors without subset-specific cytokine supplementation [18]. A recent study by Dijkstra et al. used autologous tumor organoids to enrich tumor-reactive T cells from peripheral blood, exemplifying organoid's capacity for supporting T cell maturation and activation in a tumor-specific manner [19]. Schafer et al. developed an in vivo xenotransplantation approach allowing for the study of functionally mature human microglia within a physiologically relevant human brain organoid model [20]. Finally, Neal et al. developed an air-liquid interface (ALI) method that propagates patient-derived organoids (PDOs) with native embedded immune cells [3]. They reported that this approach allows for the preservation of the original tumor T cell receptor (TCR) spectrum, potentially providing a more accurate representation of the tumor immune clonal dynamics and evolution [3].

Together, these studies underscore the growing utility of organoid technology as a powerful platform for modeling immune cell development within tissue-specific contexts. By recapitulating key aspects of the native microenvironment, organoids enable more physiologically relevant investigations into immune responses, disease progression, and therapeutic strategies than 2D cultures.

3.1.4 Modeling Inflammatory Pathophysiology

Beyond cancer, organoids have been used to study the immune system in the context of various inflammatory diseases, such as inflammatory bowel disease (IBD) and asthma. By manipulating the immune cells within organoid models, researchers can observe how immune responses contribute to disease development, progression, and resolution.

Eight recent studies on inflammatory bowel disease models show that three-dimensional systems—organoids and organ-on-chip platforms—better recapitulate intestinal structure and function than two-dimensional cultures [21–28]. In these reports, 3D models reliably form architectures that mimic native intestinal tissues and capture key pathological features, including morphological abnormalities, epithelial barrier function, and inflammatory cytokine expression [22, 28]. Ojo et al. noted that organoids maintain greater genetic stability and differentiation capacity than do 2D cultures [25]. Models reporting quantitative metrics demonstrated replication of features such as barrier permeability and chemokine as well as tight junction protein expression [21, 27].

Asthma offers another inflammatory disease context for the use of organoids to study the immune system. Three-dimensional organoid models elucidate several mechanisms of asthma pathogenesis that remain obscured in conventional 2D cultures [29]. In these 3D systems, cyclic mechanical stretch enhances the alignment of nuclei and actin stress fibers in airway smooth muscle cells and increases expression of contractile proteins [29]. Mechanical stress also triggers airway constriction and discordant cyclooxygenase gene expression via YAP signaling, linking physical forces to biochemical responses [30]. Studies further document that co-culturing enables spatial reorganization of key cell types [31]. Mast cells shift from the submucosa to the airway epithelium and, together with fibroblasts and immune cells, establish a feedforward loop through IL-33 that underpins type-2 inflammation and airway remodeling [31]. In parallel, bronchial epithelial spheroid models reveal that exposure to IL-13 impairs barrier development and favors a goblet cell–dominated phenotype, accurately recapitulating asthmatic epithelial dysfunction [32].

These findings indicate that 3D models better replicate extracellular matrix complexity, mechanochemical coupling, and multidimensional cell interactions than their 2D counterparts, thereby providing enhanced insights into the molecular and cellular drivers of IBD and asthma, just to name two of the many inflammatory conditions studied with organoids.

3.2 Personalized Medicine

Beyond acting as a model for advances in basic science research, organoid models may also have potential to be used for personalizing clinical treatment and practice. Patient-specific organoids can be created from biopsies, enabling researchers to study how an individual's immune system might respond to various treatments. These personalized organoid models can be used to test immunotherapies, vaccines, or other immune-modulating drugs in a way that reflects the genetic and immune landscape of the patient, offering more tailored therapeutic options.

A tumor organoid/lymphocyte co-culture approach has been described that enables the expansion and analysis of tumor-specific T cell responses [33], with the author suggesting that this method provides a patient-specific platform for studying T cell interactions with tumor cells. Underscoring this assertion, previously mentioned studies have found that organoids retain the in vivo TCR repertoire [3] and tumor microenvironmental components [1]. These findings also outline the potential of patient-derived organoids for offering a platform for studying immune-cancer interactions in a patient- and even tumor-specific manner.

Establishing a workflow pipeline for isolating patient biopsies, generating organoids, and screening various therapeutic options is a highly active area of research, and has the potential to revolutionize clinical practice, especially for cancer patients. 2D models are less feasible for use in this context, as they require additional action to become viable in vitro, ultimately confounding any interpretation of data (Fig. 3.1).

3.2.1 Drug Screening

Organoids provide a platform for high-throughput drug screening, enabling researchers to test how different immune-modulating drugs affect immune responses in the context of a more realistic tissue structure. This approach has particularly high potential for testing new immunosuppressive or immune-enhancing agents, in addition to new combinations of existing agents in use with either old and new agents.

Astashkina et al. report that kidney organoids display toxicity biomarkers and responses to nephrotoxic agents that closely reflect in vivo endpoints, whereas flat cultures lack these clinical characteristics [34]. In cancer models, Jacobi et al. observe that organoid systems establish intercellular communication, oxygen and

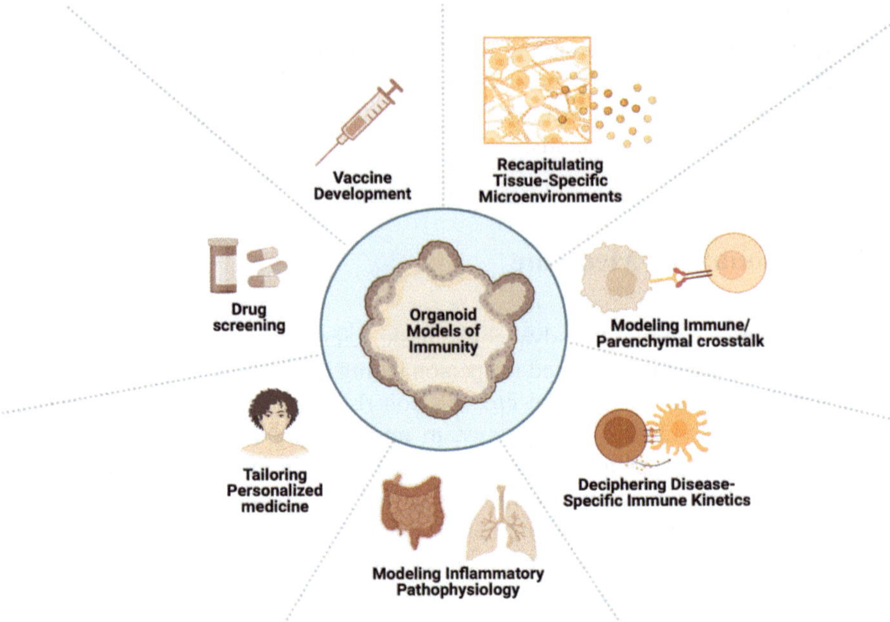

Fig. 3.1 Illustration of the key areas of preclinical and clinical research enabled by organoid-immune co-culture models

nutrient gradients, and appropriate cell polarity; these features yield drug sensitivity profiles—such as for EGFR inhibitors—that mirror clinical outcomes [35]. Hou et al. further demonstrate that bio-printed pancreatic cancer organoids, adapted for high-throughput screening in standard plates, reveal drug resistance mechanisms aligned with in vivo conditions, as evidenced by the screening of approximately 3300 approved compounds [36]. Skardal et al. extend these findings by showing that liver-tumor organoids and other 3D platforms capture complex drug-pathway interactions and maintain in vivo-like phenotypes [37].

In summary, the studies indicate that organoid models better recapitulate in vivo cellular organization, microenvironmental gradients, and drug response patterns, thereby offering enhanced predictive accuracy for drug screening compared to traditional 2D cultures.

3.3 Vaccine Development

Organoids, particularly those that model mucosal tissues like the gut or respiratory tract, can be used to test vaccine responses. These models can mimic how vaccines trigger immune responses at mucosal surfaces and provide insights into vaccine efficacy and potential side effects even before clinical trials. Vaccination against a wide range of pathologies, from viruses to cancer, has been studied with organoids.

With regard to viruses, organoids have been shown to be capable of elucidating virus–host receptor interactions and evaluate the efficacy of neutralizing antibodies against SARS-CoV-2 in organoids derived from, among others, the lung, kidney, and intestine [38]. Hong and Seo demonstrated that human intestinal organoids yield viral replication—up to ten times higher for rotavirus—offering improved culture systems for vaccine strain production [39]. Wagar proposed that human tonsil organoids accurately capture immunogenicity and immunodominance, aiding antigen selection [40]. Showcasing the flexibility of organoid systems to develop vaccines against viral or bacterial infections, Moeller et al. used a three-dimensional B cell follicular organoid to simulate germinal center responses that correlate with in vivo humoral immunity, facilitating rapid antibody discovery and conjugate vaccine efficacy prediction [41].

Organoid-immune co-cultures can also be used when developing cancer vaccines [42]. For instance, patient-derived organoids co-cultured with autologous immune cells—such as tumor-infiltrating lymphocytes and peripheral blood mononuclear cells—have been employed to assess cytokine secretion, T cell activation, and cytotoxicity [42, 43]. Platforms such as these have demonstrated the capacity for high-throughput, multiplexed screening, and have successfully identified immunogenic targets, including epigenetic regulators and novel checkpoint molecules [42, 44]. Notably, one study reported and 85% concordance with patient outcomes [45]. By contrast, more traditional approaches generally only allow moderate throughput and, as already discussed, offer a limited ability to model the complexity of immune-tumor interactions [42]. Collectively, these findings demonstrate that organoid immune co-culture systems offer a more physiologically relevant and efficient strategy for identifying immunogenic targets for the development of cancer vaccines.

3.4 Conclusion

Organoid models have emerged as transformative tools in immunological research, bridging the gap between simplistic 2D cultures and complex in vivo systems. Their ability to faithfully replicate tissue architecture, preserve immune microenvironments, and allow for patient-specific modeling has positioned them at the forefront of studies in immune interaction, inflammatory disease, immune development, and therapeutic response. From enabling deeper understanding of immune mechanisms to advancing drug and vaccine development, organoids offer unparalleled physiological relevance and experimental flexibility. As technologies advance, organoid platforms are likely to play an increasingly vital role in both fundamental immunological discoveries and personalized clinical applications.

References

1. Salahudeen AA, Zhu J, Jihang J, Batish A, Sutha K, Neal J, Giangarra V, Montesclaros L, Sapida J, Sharifi O, Lee J, Zheng G, Wagh D, Coller J, Neal J, Padda S, Sabatti C, Kuo C. Abstract 987: organoid-based characterization of patient tumors and microenvironments at single cell resolution. Cancer Res. 2018;78:987.
2. Bar-Ephraim YE, Kretzschmar K, Clevers H. Organoids in immunological research. Nat Rev Immunol. 2020;20(5):279–93.
3. Neal JT, et al. Organoid modeling of the tumor immune microenvironment. Cell. 2018;175(7):1972–1988.e16.
4. Panciera T, et al. Reprogramming normal cells into tumour precursors requires ECM stiffness and oncogene-mediated changes of cell mechanical properties. Nat Mater. 2020;19(7):797–806.
5. Piersma B, Hayward MK, Weaver VM. Fibrosis and cancer: a strained relationship. Biochim Biophys Acta Rev Cancer. 2020;1873(2):188356.
6. Wang Y, et al. Matrix stiffness regulates macrophage polarization in atherosclerosis. Pharmacol Res. 2022;179:106236.
7. Vlachogiannis G, et al. Patient-derived organoids model treatment response of metastatic gastrointestinal cancers. Science. 2018;359(6378):920–6.
8. Drost J, Clevers H. Organoids in cancer research. Nat Rev Cancer. 2018;18(7):407–18.
9. Chan IS, Ewald AJ. Organoid co-culture methods to capture cancer cell-natural killer cell interactions. Methods Mol Biol. 2022;2463:235–50.
10. Roper J, Shen X. Novel three-dimensional cultures of patient-derived cancer and tumor immune cells. Gastroenterology. 2019;157(1):260–1.
11. Scognamiglio G, De Chiara A, Parafioriti A, Armiraglio E, Fazioli F, Gallo M, Aversa L, Camerlingo R, Cacciatore F, Colella G, Pili R, de Nigris F. Patient-derived organoids as a potential model to predict response to PD-1/PD-L1 checkpoint inhibitors. Br J Cancer. 2019;121:979–82.
12. Zhang H, Zhou J, Zhu S, Zheng J, Shang L, Li C, Yan X, Zhang R, Zang M, An AX, Xiaoxi X, Wang S, Li H, Huang Y. 527 Tumor organoid and immune cell co-culture system potentiates immuno-oncology drug development. J Immunother Cancer. 2020;8:A322–3.
13. Kolaczkowska E, Kubes P. Neutrophil recruitment and function in health and inflammation. Nat Rev Immunol. 2013;13(3):159–75.
14. Barbul A, Regan MC. Immune involvement in wound healing. Otolaryngol Clin N Am. 1995;28(5):955–68.
15. Hu C, et al. Dissecting the microenvironment around biosynthetic scaffolds in murine skin wound healing. Sci Adv. 2021;7(22):eabf0787.
16. Cavagnero KJ, Gallo RL. Essential immune functions of fibroblasts in innate host defense. Front Immunol. 2022;13:1058862.
17. Tsoumakidou M. The advent of immune stimulating CAFs in cancer. Nat Rev Cancer. 2023;23(4):258–69.
18. Jowett GM, et al. Organoids capture tissue-specific innate lymphoid cell development in mice and humans. Cell Rep. 2022;40(9):111281.
19. Dijkstra KK. Abstract IA001: patient-derived tumor organoids to probe the cancer immune interface. Mol Cancer Ther. 2023;22:IA001.
20. Schafer ST, et al. An in vivo neuroimmune organoid model to study human microglia phenotypes. Cell. 2023;186(10):2111–2126.e20.
21. O'Connell L, Winter DC, Aherne CM. The role of organoids as a novel platform for modeling of inflammatory bowel disease. Front Pediatr. 2021;9:624045.
22. Di Giorgio C, et al. Modeling inflammatory bowel disease by intestinal organoids. Recent Adv Inflamm Allergy Drug Discov. 2023;17(1):39–53.
23. Dotti I, Salas A. Potential use of human stem cell-derived intestinal organoids to study inflammatory bowel diseases. Inflamm Bowel Dis. 2018;24(12):2501–9.

24. Kaden T, et al. Leveraging organ-on-chip models to investigate host-microbiota dynamics and targeted therapies for inflammatory bowel disease. Adv Healthc Mater. 2025;14(10):e2402756.
25. Ojo BA, VanDussen KL, Rosen MJ. The promise of patient-derived colon organoids to model ulcerative colitis. Inflamm Bowel Dis. 2022;28(2):299–308.
26. Park SE, et al. Geometric engineering of organoid culture for enhanced organogenesis in a dish. Nat Methods. 2022;19(11):1449–60.
27. Rallabandi HR, et al. Evaluation of intestinal epithelial barrier function in inflammatory bowel diseases using murine intestinal organoids. Tissue Eng Regen Med. 2020;17(5):641–50.
28. d'Aldebert E, et al. Characterization of human colon organoids from inflammatory bowel disease patients. Front Cell Dev Biol. 2020;8:363.
29. Asano S, et al. Cyclic stretch enhances reorientation and differentiation of 3-D culture model of human airway smooth muscle. Biochem Biophys Rep. 2018;16:32–8.
30. Kilic O, et al. A microphysiological model of the bronchial airways reveals the interplay of mechanical and biochemical signals in bronchospasm. Nat Biomed Eng. 2019;3(7):532–44.
31. Altman MC, et al. Airway epithelium-shifted mast cell infiltration regulates asthmatic inflammation via IL-33 signaling. J Clin Invest. 2019;129(11):4979–91.
32. Pat Y, et al. Differentiation of bronchial epithelial spheroids in the presence of IL-13 recapitulates characteristic features of asthmatic airway epithelia. Allergy. 2022;77(7):2229–33.
33. Sahin U. Studying tumor-ReacTive T cells: a personalized organoid model. Cell Stem Cell. 2018;23(3):318–9.
34. Astashkina AI, et al. Comparing predictive drug nephrotoxicity biomarkers in kidney 3-D primary organoid culture and immortalized cell lines. Biomaterials. 2012;33(18):4712–21.
35. Jacobi N, et al. Organotypic three-dimensional cancer cell cultures mirror drug responses in vivo: lessons learned from the inhibition of EGFR signaling. Oncotarget. 2017;8(64):107423–40.
36. Hou S, et al. Advanced development of primary pancreatic organoid tumor models for high-throughput phenotypic drug screening. SLAS Discov. 2018;23(6):574–84.
37. Skardal A, et al. Liver-tumor hybrid organoids for modeling tumor growth and drug response in vitro. Ann Biomed Eng. 2015;43(10):2361–73.
38. Chen KG, Park K, Spence JR. Studying SARS-CoV-2 infectivity and therapeutic responses with complex organoids. Nat Cell Biol. 2021;23(8):822–33.
39. Hong KJ, Seo SH. Organoid as a culture system for viral vaccine strains. Clin Exp Vaccine Res. 2018;7(2):145–8.
40. Wagar L. Small centers of defense. Science. 2022;375(6583):830.
41. Moeller TD, et al. Profiling germinal center-like B cell responses to conjugate vaccines using synthetic immune organoids. ACS Cent Sci. 2023;9(4):787–804.
42. Zhou Z, et al. An organoid-based screen for epigenetic inhibitors that stimulate antigen presentation and potentiate T-cell-mediated cytotoxicity. Nat Biomed Eng. 2021;5(11):1320–35.
43. Magre L, et al. Emerging organoid-immune co-culture models for cancer research: from oncoimmunology to personalized immunotherapies. J Immunother Cancer. 2023;11(5):e006290.
44. Zhou Z, et al. A T cell-engaging tumor organoid platform for pancreatic cancer immunotherapy. Adv Sci (Weinh). 2023;10(23):e2300548.
45. Votanopoulos KI, et al. Model of patient-specific immune-enhanced organoids for immunotherapy screening: feasibility study. Ann Surg Oncol. 2020;27(6):1956–67.

Chapter 4
Lymphoma Organoid Models for Evaluating and Targeting Tumor Immune Microenvironment

Rohat Esmer, Bahadır Köylü, and Şahin Laçin

Abstract Lymphoma organoid models are very powerful new tools. They excel at recapitulating certain patient-specific features—for example, the interplay of a patient's T cells with their tumor B cells—and thereby offer unprecedented opportunities for in vitro immunotherapy trials on patient tissue. However, current models still fall short of recreating the full complexity of the lymphoma microenvironment, including the circulatory system, a complete immune repertoire, and long-term tumor–immune dynamics. Ongoing improvements aim to extend immune cell longevity, include additional immune and stromal components, and standardize protocols to reduce technical variability. As these challenges are addressed, lymphoma organoids are expected to become increasingly reliable and predictive. Already, their success in modeling patient responses is encouraging, and careful validation is needed. Ultimately, the integration of organoid assays into preclinical and clinical workflows will require a balanced understanding of their limitations—using organoids as complementary to clinical judgment and genomic biomarkers, rather than infallible stand-alone predictors. With continued refinement, lymphoma organoid co-cultures hold great promise to bridge the gap between laboratory testing and real-world patient outcomes, accelerating the development of truly personalized cancer therapies.

Keywords Lymphoma organoids · Personalized cancer therapies · Preclinical models

R. Esmer
Koç University School of Medicine, Istanbul, Turkey
e-mail: resmer18@ku.edu.tr

B. Köylü · Ş. Laçin (✉)
Department of Medical Oncology, Koç University School of Medicine, Istanbul, Turkey
e-mail: bkoylu@kuh.ku.edu.tr

O. S. Somuncu (ed.), *Organoids for Preclinical Disease Modeling and Personalized Medicine*, Stem Cell Biology and Regenerative Medicine 78,
https://doi.org/10.1007/978-3-032-10024-5_4

4.1 Introduction

Three-dimensional (3D) patient-derived lymphoma organoids (PDLOs or "lymphoma tumoroids") are emerging as advanced in vitro models that preserve the complex cellular makeup of lymphomas [1, 2]. Unlike two-dimensional (2D) cultures, which often fail to sustain primary lymphoma cells, PDLOs provide *a microenvironment-rich 3D context* that better mimics lymphoid tissue architecture [1, 3, 4]. Recent studies underscore that lymphoma cell survival in vitro depends on faithfully recapitulating the tumor niche: the myriad interactions with immune and stromal cells and extracellular matrix (ECM) that tumor cells rely on in patients [5, 6]. *PDLO technology bridges this gap*, allowing patient biopsy–derived cells to self-organize or be co-cultured in 3D matrices, thereby maintaining viability and phenotype for experimental therapy testing [1, 2, 7]. This methodology is particularly valuable given the heterogeneity of lymphomas.

4.2 Establishing Lymphoma Organoid Models

4.2.1 Methodologies for Developing Patient-Derived Lymphoma Organoids (PDLOs)

Patient tissue procurement and processing are the first critical steps in PDLO establishment. Excisional lymph node biopsies or tumor resections are preferred to obtain an intact repertoire of malignant and supportive cells; indeed, using sufficient tumor tissue is key to represent the full complexity of the tumor microenvironment (TME) [2, 7]. For instance, organoids have been successfully generated from cryopreserved diagnostic and relapse lymph node specimens from patients with follicular lymphoma (FL), including those with high-grade transformation to diffuse large B cell lymphoma (DLBCL) [2]. Kastenschmidt et al. cultured $n = 12$ FL patient samples as organoids, demonstrating *robust outgrowth* and preservation of the original cellular constituents over several passages [8]. Similarly, Santamaría-Martínez et al. collected 27 patient tumors spanning multiple lymphoma subtypes and established small ex vivo tissue fragments ("lymphomoids") that retained native histology and cell composition [9]. These intact tissue organoids could be maintained for up to a week in vitro, enabling short-term drug response assays that correlated with clinical outcomes. Such findings highlight that *high success rates can be achieved* when ample viable tumor and microenvironment cells are available [2, 7]. By contrast, smaller needle biopsy samples often yield limited cell numbers and may under-represent intra-tumoral heterogeneity, posing a challenge for organoid initiation [1, 2]. To maximize viability, gentle enzymatic or mechanical dissociation protocols are used to free lymphoma cells from stroma while keeping key immune infiltrates intact.

3D scaffold selection is another pivotal consideration in PDLO methodologies [10]. Lymphomas lack a discrete anatomical structure, so unlike epithelial organoids, they do not *self-organize* into organ-like cysts and often require an artificial matrix to assemble into 3D micro-tissues [9, 10]. Early efforts used hydrogel matrices tailored to lymphoid cells—for example, methylcellulose or alginate hydrogels—to support lymphoma spheroid formation [10, 11]. More recently, *collagen I matrices* have been widely adopted to provide an analog of lymph node interstitial tissue. In a DLBCL model by Foxall et al., patient-derived lymphoma cells were embedded in a type I collagen hydrogel, which reliably yielded multicellular spheroids harboring tumor B cells intermingled with support cells [12]. The collagen scaffold maintained the *3D architecture and polarization* of lymphoma clusters, better reflecting in vivo growth than liquid culture. Notably, lymphoma organoids in collagen showed cell–cell and cell–ECM contacts that reproduced those in lymphoid tissue, and exhibited hypoxic gradients and nutrient diffusion kinetics closer to tumors [2, 12]. Synthetic *engineered hydrogels* functionalized with specific ligands have further refined the microenvironment. For example, integrin-specific PEG-based hydrogels presenting lymphoma-associated adhesion motifs were shown to enhance the growth and drug resistance of aggressive B cell lymphoma cells in vitro by engaging B cell receptor and integrin signaling pathways [13, 14]. Similarly, Bahlmann et al. recently designed a biomimetic ECM-mimicking cryogel (comprised of hyaluronan, gelatin, and fibronectin with adhesive peptides) to model the nodal niche in Hodgkin lymphoma; this scaffold successfully induced *tumor-associated macrophage (TAM) infiltration* into the 3D matrix, recapitulating a key stromal interaction of the TME [15]. These examples illustrate how matrix choice can be tuned to different lymphoma contexts: a simple collagen gel may suffice to aggregate cells, whereas *custom hydrogels* can incorporate biochemical cues (e.g., integrin ligands, chemokines) to actively support lymphoma subtypes with particular microenvironmental dependencies. In practice, Matrigel (a basement membrane extract) is less commonly used for lymphoma organoids, as lymphoid cells do not require an epithelial basement membrane; instead, neutral scaffolds like collagen or semi-solid methylcellulose provide structure without unwanted differentiation signals [9, 10]. The physical properties of the matrix (stiffness, porosity) also affect outcomes: a compliant matrix can facilitate cell migration and realistic spatial organization, whereas an overly rigid or dense matrix might impede nutrient diffusion or immune cell motility [2, 10, 11].

Co-culture systems and media formulation are meticulously optimized in PDLO protocols to recreate the cellular crosstalk of the lymphoma niche [10, 16]. Lymphomas are infiltrated by an array of non-malignant cells—T cells, macrophages, dendritic cells, fibroblastic reticular cells, etc.—which provide survival signals [17]. Whenever possible, PDLOs aim to maintain these *autologous infiltrating cells* from the patient tumor [8, 16]. In FL organoids reported by Kastenschmidt et al., the dissociated tumor cell suspension inherently contained the patient's own T cells, macrophages, and stromal elements, so no exogenous cell types were needed to sustain the culture [8]. In other cases, supplemental feeder cells or immune cells are deliberately added [10, 12]. Foxall et al. demonstrated that introducing "lymphoid stromal" fibroblasts (derived by differentiating adipose stem cells) and monocyte-derived macrophages into DLBCL cultures dramatically improved in vitro lymphoma cell survival, whereas lymphoma cells alone underwent apoptosis [12]. This *3D tri-culture model* (lymphoma + fibroblasts + macrophages) better resembled the tumor's cellular ecosystem and was leveraged to study antibody therapies (e.g., rituximab-mediated phagocytosis by macrophages) under more physiological conditions. The composition of co-culture is carefully chosen to match the lymphoma subtype: for instance, mantle cell lymphoma (MCL) organoids have been built by co-culturing primary MCL B cells with *autologous T cells* (from the same patient's blood or node) and *allogeneic monocytes* from healthy donors, which differentiate into M2-like lymphoma-associated macrophages in situ [18]. This recapitulates MCL's pro-tumoral TME signals (e.g., M2 macrophages and T cell exhaustion) at the cost of introducing non-self-myeloid cells. Indeed, one drawback of using donor feeder cells is the potential for *alloreactivity* or altered signaling. Ideally, fully autologous PDLO models are pursued [8]. Achieving that often requires supplementing the medium with cytokines and growth factors to replace helper cell signals. For example, to establish FL spheroids without an external feeder layer, Faria et al. designed an "*LN cytokine cocktail*" containing recombinant CD40 ligand (to mimic T follicular helper cell contact), interleukin-4 and -21 (germinal center T cell cytokines that promote B cell growth), and IL-15 (to support T and NK cells) [16]. FL cells cultured with this cocktail (plus autologous T cells and donor monocytes) spontaneously formed 3D spheroids (FL-PDLS) and maintained high viability ex vivo. Notably, these organoids contained ~60% tumor B cells and ~ 15% T cells, closely mirroring the original tumor's cellular makeup.

Other media additives commonly include B cell growth factors (BAFF/BLYS, IL-6, IL-10), T cell stimulants (IL-2 for CTLs or IL-21 for T_{fh}), and even *stromal cell-conditioned media* to provide a baseline level of chemokines [16, 17]. Through iterative testing, each group optimizes a recipe that best supports their lymphoma subtype of interest [10, 16]. Taken together, these methodology advances—proper biopsy sourcing, 3D matrices, co-culture with representative cell types, and enriched media—have greatly increased the *success rates* of PDLO generation across lymphoma subtypes. Indolent lymphomas like FL and MCL have shown particularly high take rates, as their malignant cells are highly dependent on the niche signals that organoids strive to preserve [9, 16, 18]. Even aggressive lymphomas (DLBCL, Burkitt) can be propagated as organoids when provided sufficient stromal/immune support, although they may be more finicky [9, 12]. Impressively, one study was able to establish ex vivo 3D cultures from refractory DLBCL patients and accurately predict their lack of response to various targeted drugs, demonstrating that *patient-derived lymphoid organoids can be created from clinically high-risk tumors* and retain their therapy-resistant phenotype in vitro [9]. In contrast, classic Hodgkin lymphoma (cHL) remains challenging for organoid modeling—the malignant Reed–Sternberg cells constitute <1% of the tumor and rely on a vast reactive infiltrate [17, 19].

4.3 Preserving the Native Tumor Microenvironment in Organoid Cultures

A defining goal of lymphoma organoid models is to preserve the native TME as intact as possible [8, 16]. The TME in lymphomas plays an active role in disease behavior [20]. Recent patient-derived organoids such as *FL PDLOs* were shown to retain a complex mixture of autologous T cells, B cells, macrophages, and dendritic cells in culture, closely matching the cell type proportions of the original lymph node biopsy [8, 16]. Kastenschmidt et al. reported that the TME cellular composition in FL organoids remained stable over at least 21 days, with minimal loss of minority cell populations [8]. Phenotypic analysis confirmed that key T cell subsets present in vivo—including PD-1^{+} T_{fh} cells, regulatory T cells, and exhausted $CD8^{+}$ T cells—were all preserved in the organoids. Likewise, *MCL spheroids* developed by Araujo-Ayala et al. contained the full immune milieu characteristic of MCL lymph nodes: the malignant B cells proliferated in proximity to $CD4^{+}$ and $CD8^{+}$ T cells (which gradually upregulated exhaustion markers TIM-3, TIGIT, PD-1), while added monocytes differentiated into $CD163^{+}$ M2-like macrophages [18]. The fact that these cultures naturally reproduced T cell exhaustion and immunosuppressive macrophage polarization is strong evidence that the *spatial and paracrine signals* of the native TME were effectively reconstituted [8, 18]. This is critical because many modern therapies (checkpoint inhibitors, bispecific T cell engagers, CAR-T cells) act on the tumor–immune interface [20]. Only by preserving that interface can organoids reliably predict responses to such immunotherapies [9, 16]. Indeed, in FL-PDLS models that included autologous T cells, treatment with T cell–engaging bispecific antibodies (anti-CD3:CD20 or CD3:CD19) led to vigorous in situ T cell activation and lymphoma cell killing [16]. Interestingly, the degree of tumor cell clearance in those organoids varied with the immunological makeup: organoids with high starting T_{fh} cell content were more resistant to T cell–engagers, consistent with T_{fh}-driven immunosuppressive loops observed in patients [8, 16]. Such subtle TME-dependent effects could only be captured in an *autologous* system where patient T cells and lymphoma cells coexist. By contrast, if one were to use lymphoma cells alone or mixed with allogeneic immune cells, these nuanced interactions (e.g., T_{fh} helping lymphoma cells evade T_{killer} cells) might be missed. Thus, *maintaining the native immune context* within organoids is paramount. Empirically, studies have shown that PDLOs recapitulate not just the cellular composition but also the gene expression profiles of the tumor and immune compartments [9, 16, 18]. RNA sequencing (RNA-seq) of FL organoids revealed enrichment of the same B cell receptor signaling, cell cycle, and mTOR pathways as in the original tumors, and T cells in organoids displayed upregulation of exhaustion markers (PD-1, LAG-3, TIGIT) analogous to T cells isolated from lymphoma patients [8, 16]. Similarly, MCL organoids reflected the lymphoma's oncogenic NF-κB and BCR signaling signatures, alongside interferon-gamma and granzyme B production when immune cells were engaged by therapy [18]. *These results validate that PDLOs can preserve the functional TME,* enabling investigators to probe how the native context contributes to drug responses. Preserving the native TME allows *identification of TME biomarkers of response* (like T cell exhaustion or T_{fh} abundance) and enables testing immunotherapies in a setting where target and effector cells interact as they do in the patient [16, 21].

Organoids strive to incorporate the *stromal and structural elements* of the lymphoma niche [10]. Preserving this tissue architecture is vital for modeling processes like immune cell trafficking, antigen presentation, and hypoxia [20]. Some PDLO approaches use an *intact tissue fragment* strategy to inherently retain native spatial architecture [9, 10]. For dissociated-cell organoids, an alternative is to *add back key stromal components* [12, 16]. For example, the 3D DLBCL model contains seeded lymphoma cells onto a layer of human mesenchymal stromal cells within a collagen matrix [12]. Stromal cells serve as a feeder layer—expressing adhesion molecules and secreting cytokines like IL-6—that the lymphoma cells could latch onto, mimicking their residence in lymph node niches. The presence of stromal fibroblasts also provides a surface for macrophages and T cells to interact with, effectively recreating the *cellular triad* (tumor–fibroblast–macrophage) observed in vivo. However, the source of stromal cells can influence the fidelity of the model [10, 21]. In some instances, researchers have used *xenogeneic feeder cells* (e.g., murine fibroblast cell lines) to support human lymphoma organoids, which aids lymphoma cell growth but may distort the immune microenvironment [10].

Lymphomas often remodel their ECM [17]. To mirror this, Bahlmann et al.'s cHL model included fibronectin in the cryogel scaffold, and they observed macrophages actively migrating into the fibronectin-rich matrix in the presence of Hodgkin Reed–Sternberg cells [15]. The invading TAMs in their model expressed matrix metalloproteinases (MMPs) to burrow through the gel, akin to how TAMs infiltrate tumor stroma in patients. Interestingly, when MMP inhibitors (e.g., marimastat, batimastat) were added, TAM infiltration into the HL organoid was blocked, validating the model as a tool to study cell–ECM dynamics and revealing MMPs as potential therapeutic targets to disrupt the TME in HL. This exemplifies how spatial architecture and ECM components in organoids permit functional readouts (like cell invasion) that are impossible to capture in suspension culture.

In current PDLOs, common practice is to incorporate at least a rudimentary ECM scaffold to provide structure, and to allow stromal cells to produce additional matrix proteins over time [10, 16]. The ex vivo tumor slice cultures can preserve native architecture and cell interactions for a few days. These have been used to test drugs on actual patient tissue directly, maintaining TME signaling during the short culture. Santamaría-Martínez et al.'s lymphomoid approach is essentially a variant of slice culture (small minced fragments cultured at an air–liquid interface), and it retained not only cellular composition but also the *spatial arrangement* of tumor cells and immune cells within the fragment, as confirmed by immunohistochemistry and spatial transcriptomics [9]. The trade-off is that tissue fragments lack longevity (viable for ~3–7 days) and are not easily scalable or manipulable like dissociated-cell organoids [9, 10]. Nonetheless, they serve as a gold standard for TME preservation, since the structure is initially intact.

4.4 Challenges and Solutions in Maintaining Cellular Heterogeneity and Viability over Time

Despite their promise, lymphoma organoids face several challenges in long-term maintenance, notably in preserving the full cellular heterogeneity of the tumor and sustaining viability without drift or decay [10]. One challenge is clonal selection and evolution in culture [22]. Lymphomas often contain multiple subclones or diverse cell subpopulations; when removed from the patient and grown ex vivo, certain clones may have a growth advantage and overtake others [9, 20]. This can lead to an organoid that no longer reflects the original tumor's heterogeneity. Investigators have observed that inadequate co-culture conditions can cause selective loss of specific immune subsets or lymphoma subclones [8, 9, 16].

Achieving an equilibrium that "suits all cell types involved" is inherently difficult [10]. In practice, groups like Kastenschmidt et al. have countered this by maintaining organoids in high-density, small-volume formats (e.g., 96-well U-bottom plates) that encourage cell–cell contacts and minimize clonal isolation [8]. Their FL PDLOs remarkably preserved the original clonal architecture. Nonetheless, this was achieved in an indolent lymphoma; more aggressive lymphomas may not be as genetically or phenotypically stable ex vivo [9, 16].

Genomic instability can be a concern for prolonged cultures [10, 22]. So far, short-term organoids (2–4 weeks) seem to largely maintain the input tumor's mutational profile but longer passages could allow genetic drift [9, 22]. Most PDLO studies to date actually avoid continuous passaging to prevent this; instead, they generate organoids fresh from cryopreserved cells for each experiment, treating them as *patient avatars* rather than perpetual cell lines [9, 16].

Another challenge is cellular senescence and exhaustion over time [10]. Notably, T cells in organoids can become exhausted without periodic antigenic stimulation or fresh influx of cytokines. Faria et al. observed that by day 7 of FL-PDLS culture, the autologous T cells expressed higher levels of exhaustion markers (PD-1, LAG-3, TIM-3) than at initiation, and their functional capacity to kill tumor cells started to wane [16]. Similarly, stromal cells that are not immortalized will eventually stop proliferating. Primary fibroblasts can undergo culture-induced senescence, and monocyte-derived macrophages may have limited survival (often a few weeks) before dying or polarizing terminally [10].

To surmount these challenges, researchers are developing innovative culture techniques and bioreactor systems to better sustain organoids long term. One promising approach is the use of air–liquid interface (ALI) cultures, where organoids are grown on a porous membrane at an air–media interface rather than fully submerged [9, 23]. The ALI method provides superior oxygenation and allows secreted growth factors to form gradients as they would in tissue. Neal and colleagues successfully propagated organoids from over 100 primary tumors for extended periods using ALI, maintaining not only tumor cells but also tumor-infiltrating lymphocytes and myeloid cells in the cultures [23]. Applying ALI to lymphoma organoids could similarly improve viability: the enhanced oxygen and nutrient delivery can delay necrosis, and the semi-dry surface may foster the formation of lymphoid aggregates with distinct zones. Santamaría-Martínez's lymphomoid method effectively used an ALI setup to keep slices alive, and one can envision hybrid techniques where dissociated lymphoma cells are re-seeded onto an ALI platform to prolong their survival ex vivo [9].

Another strategy is to employ dynamic perfusion and suspension bioreactors [24]. Gentle rocking or perfusion of media can prevent waste buildup and better distribute nutrients in 3D cultures. In one example, CLL cells co-cultured with bone marrow stromal cells on a sponge scaffold were placed in a rotating bioreactor; the continual flow of media and physical rotation supported leukemia–stroma contact and maintained drug-resistant subpopulations that would die in static culture [25]. Another solution to maintain heterogeneity is to bank and pool multiple organoids from the same patient sample [9, 16]. Because each organoid in a well might stochastically favor certain clones, pooling replicates or seeding at high density can average out these effects [10, 16]. High-throughput platforms allow generation of dozens of mini-organoids from one sample [26]. This not only yields replicates for experimentation but also preserves heterogeneity: sequencing data can be aggregated across organoids to ensure all sub-clonal lineages are represented [9, 22]. Indeed, PDLO protocols often emphasize using multiple organoid replicates per condition to capture the range of tumor responses [10, 16].

Finally, it is worth noting that maintaining viability is not solely a technical hurdle but also a biological one: lymphoma cells exited from the human body may undergo phenotypic changes as they adapt to culture [9, 10, 22]. To mitigate this, some groups have kept organoids in conditions as close to the human body as possible [10, 16]. Others have co-transplanted organoids into immunodeficient mice ("organoid xenografts") to periodically rejuvenate them in vivo, then re-isolated them [27]. As extreme as that sounds, it has been piloted in solid tumors and might be conceivable for lymphomas as well.

4.5 Evaluating the Immune Microenvironment in Lymphoma Organoids

In recent years, high-throughput technologies have transformed our comprehension of the TME [28]. However, animal models and conventional 2D in vitro cell cultures have significant limitations that prevent them from fully replicating the complex, human-specific biological processes [29]. The introduction of 3D cultures marked a significant breakthrough in oncological research [30]. Tumor organoids, which are a specific subset of 3D cultures derived from patient-specific tumor samples, provide an invaluable platform for understanding the TME, investigating tumorigenic pathways, and conducting drug screening for personalized therapeutic approaches, ultimately allowing for more tailored and effective treatment plans for individual patients [5, 27, 31]. Their application as promising preclinical models for studying a range of malignancies, including lymphoma, is rapidly expanding.

4.5.1 Methods for Assessing Immune Cell Composition and Interactions Within Lymphoma Organoids

4.5.1.1 Use of Immunohistochemistry, Flow Cytometry, and Gene Expression Profiling

Gaining deeper insights into the TME is essential for stratifying patients more effectively, predicting treatment responses through reliable biomarkers, and guiding the development of improved therapeutic strategies [20, 28]. Available methods for characterizing the TME include immunohistochemistry (IHC), flow cytometry, and gene expression profiling [32]. Conventional IHC techniques are limited in their ability to evaluate the complex TME, particularly in assessing the spatial distribution of immune cells [32]. Although sequential staining of consecutive sections can be used, this approach does not allow for high-dimensional co-expression analysis, potentially resulting in the loss of critical spatial information [28, 32]. To address these challenges, multiplex imaging technologies have been developed. These advanced methods enable multiplexed staining and standardized quantitative analysis, offering deeper insights into the spatial dynamics of the TME even in small tissue samples [28, 32]. Flow cytometry contributes significantly to the diagnosis of various hematologic malignancies, subtyping of solid tumors, and detection of metastatic disease, particularly micro-metastases in lymph nodes [20]. Gene expression profiling in tumor organoids serves multiple purposes, including validating organoid composition and functionality, assessing tumor heterogeneity, identifying novel signaling pathways and tumor-associated cell types potentially involved in tumorigenesis, and uncovering genomic regulatory elements and epigenetic modifications of chromatin [27].

While conventional methods—such as flow cytometry and IHC for analyzing tumor microenvironment composition, and real-time polymerase chain reaction (RT-PCR), microarrays, and bulk RNA sequencing for gene expression profiling—have significantly advanced our understanding of the TME, they are limited in their ability to resolve cell-type–specific information within heterogeneous populations like tumor organoids [27, 32]. To overcome these limitations and more effectively investigate the TME, advanced techniques such as mass cytometry, imaging mass cytometry, multiplex imaging technologies, single-cell RNA sequencing (scRNA-seq), spatial RNA sequencing (spRNA-seq), and artificial intelligence–based approaches have become essential tools for dissecting the cellular complexity of tumor organoids [28, 32].

4.5.1.2 Multiplex Immunohistochemistry (mIHC)/Multiplex Immunofluorescence (mIF)

Multiplex immunohistochemistry (mIHC) provides an ability to label multiple markers on a single tissue section. Two primary techniques in mIHC are multiplexed immunohistochemical consecutive staining on a single slide (MICSSS) and sequential immunoperoxidase labeling and erasing (SIMPLE), both of which facilitate sequential staining with multiple dyes using cycles of antigen labeling, image scanning, and chromogenic substrate decolorization on a single slide [28, 32]. Both methods permit the marking of only one marker in each staining round, which limits the number of markers to around 5–10 [28, 32]. Although specific applications of MICSSS, SIMPLE, and mIF in lymphoma organoid models are limited, existing studies have demonstrated promising outcomes, allowing researchers to explore previously uncharted aspects of the lymphoma TME [9, 16, 28].

4.5.1.3 Mass Cytometry (Cytometry by Time-of-Flight Mass Spectrometry/CyTOF)

Conventional flow cytometry is limited by spectral overlap between fluorescent dyes, which restricts the number of detectable markers. Mass cytometry (CyTOF) overcomes this limitation by using metal isotope-conjugated antibodies, enabling the simultaneous detection of over 40 cellular markers with minimal signal interference [28]. CyTOF improves resolution, enhances sensitivity, and facilitates the characterization of rare immune subsets, such as early activated T cells with unique metabolic profiles [28]. However, limitations include low throughput, lengthy sample preparation, lack of scatter measurements, lack of spatial information, and high costs.

4.5.1.4 Imaging Mass Cytometry (IMC)

To date, the use of imaging mass cytometry in lymphoma research has been largely confined to patient-derived tissues. However, these studies highlight the technique's potential to unravel the complex spatial architecture of the lymphoma microenvironment [28, 32]. The application of IMC to lymphoma organoid models remains unexplored and constitutes a significant frontier for future investigations.

4.5.1.5 Single-Cell RNA Sequencing (scRNA-seq) and Spatial RNA Sequencing (spRNA-seq)

RNA-seq provides a more detailed transcriptomic profile than DNA sequencing for deciphering complex cellular interactions within the TME, including both immune and non-immune compartments [27, 28]. Technological advancements have expanded RNA-seq capabilities from bulk RNA-seq to single-cell RNA-seq (scRNA-seq) and spatial RNA-seq (spRNA-seq) [28, 32, 33]. While bulk RNA-seq captures average gene expression across entire cell populations, scRNA-seq resolves expression at the single-cell level, and spRNA-seq incorporates spatial localization into transcriptomic analysis [28, 32, 33]. scRNA-seq enables transcriptomic profiling at the single-cell level by coupling whole transcriptome amplification with next-generation sequencing [28, 33]. This technique is useful for making direct comparisons between organoids and their native tissue counterparts. While scRNA-seq offers several advantages, it can result in the loss of crucial spatial and temporal information due to the lysis of cells [32]. Recent research indicates that spatial organization plays a crucial role in influencing cancer cell behavior, as several key cellular functions (e.g., growth, metastasis, cell–matrix and cell–cell interactions, intracellular signaling, drug resistance, immune escape) are driven by the properties of the entire cell population rather than by those of individual cells [28, 32]. spRNA-seq emerges as a promising technique that facilitates in-depth analysis of the tumor microenvironment and the characterization of spatial tumor heterogeneity [28, 32]. Previous research employing scRNA-seq and spRNA-seq in lymphoma organoid models has shown that these models effectively preserve the tissue structure and its components, serving as a novel ex vivo system to evaluate the impact of anti-cancer therapies on lymphoma [8, 16].

4.5.1.6 Spatial Transcriptomics

To address the loss of spatial information in high-resolution transcriptome analyses, spatial transcriptomic technologies have been developed to measure gene activity and map its precise location within tissues [28, 32]. Spatial transcriptomics can be divided into image-based and sequencing-based technologies, with subcategories based on the specific techniques used [28, 33]. Although research on spatial transcriptomics in lymphoma organoids is still in its early stages, existing studies have provided valuable insights into the spatial interactions between lymphoma cells and TME [8, 16, 32]. The integration of spatial transcriptomics with deep learning models has the potential to enhance pathological classification of lymphoma and uncover novel diagnostic, prognostic, and predictive markers [28, 32, 33].

4.5.1.7 Artificial Intelligence (AI)

With the growing complexity of organoid models that aim to replicate in vivo tissue architecture and function, analytical methods must advance accordingly. To efficiently interpret large-scale and multidimensional omics data, including transcriptomics, proteomics, and genomics, machine learning and deep learning approaches have shown significant promise in organoid research [27, 33]. Their applications range from accurate cell type classification and validation of organoid fidelity to high-throughput drug screening, paving the way for personalized precision medicine [26, 27, 33]. Additionally, artificial intelligence (AI) can help elucidate which features or feature combinations drive predictive outcomes, potentially revealing subtle biomarkers that might otherwise be missed [33]. Although studies specifically combining lymphoma organoid models with AI remain scarce, this combination represents a promising area for innovation. Incorporating AI into the analysis of lymphoma organoids could greatly enhance the precision of therapeutic response assessments and contribute to the progress of individualized cancer treatments.

Beyond omics-based methods, other advanced technologies are widely used in organoid studies to explore immune interactions [5, 34]. ELISpot, cytokine arrays, and Luminex are employed to quantify secreted immune mediators such as cytokines, chemokines, and growth factors in co-culture systems [34]. CRISPR/Cas9 (Clustered Regulatory Interspaced Short Palindromic Repeats/Cas9) screening facilitates the identification of genes involved in immune evasion or interaction, offering insights into how immune cells respond to specific genetic alterations [35]. Furthermore, live-cell imaging and time-lapse microscopy allow for real-time visualization of immune cell movement, contact, and cytotoxicity, providing a direct view of immune-organoid interactions [5, 34].

4.6 Insight Gained from Studying the Tumor Microenvironment in Lymphoma Organoid Models

To date, numerous studies have explored the use of organoids in solid tumor research, while lymphoma has been largely neglected in this regard. Hematological cancer cells exhibit complex growth behaviors, including their ability to proliferate within the stroma of tissues like lymph nodes and bone marrow [17, 18, 20]. This, along with their interactions with the immune microenvironment, complicates the development of organoid models compared to solid tumors [10]. As a result, only a limited number of studies have been published, highlighting the gap in research. Currently, there is no standardized or reproducible lymphoma organoid model that can be consistently applied across various research applications. This lack of models limits the ability to predict patient-specific drug resistance or sensitivity, thereby hindering the development of personalized therapeutic strategies.

Despite the relatively small number of studies employing organoids in lymphoma research, the findings to date have laid the groundwork for novel hypotheses and ongoing progress in this area. Kastenschmidt et al. developed a novel organoid culture method to generate PDLOs that incorporate cells from FL TME [8]. This method successfully maintained cryopreserved FL samples from multiple patients, preserving key features such as TME composition, tumor mutations, gene expression profiles, and immune receptor dynamics over 3 weeks. Additionally, they reported that treatment of PDLOs with CD3:CD19 and CD3:CD20 bispecific antibodies resulted in significant B cell cytotoxicity and T cell activation. Li et al.

developed PDLOs that effectively recapitulate the heterogeneity, diversity of cell types, and key characteristics of primary central nervous system lymphoma [36]. In this study, PDLOs demonstrated diverse responses to different drugs. Shah et al. investigated the complex interactions between malignant activated B cell-like diffuse large B cell lymphoma (ABC-DLBCL) cells and the lymphoid TME [37]. They proposed that combinatorial inhibition of multiple aberrant signaling pathways could enhance antitumor efficacy, presenting a promising strategy for improving treatment response in ABC-DLBCL. Britto et al. used synthetic hydrogel-based lymphoma organoids to investigate how lymphoid tumor microenvironment signals affect B cell receptor (BCR) signaling and histone modifications in ABC-DLBCL [6]. This research underscores the pivotal role of the lymphoid TME in regulating lymphoma cell behavior and suggests targeting these microenvironmental interactions to improve therapeutic outcomes for high-risk patients. Tian et al. found that B and T cell lymphomas rely on different integrin signaling pathways for survival in 3D environments, with follicular dendritic cells playing a key role in promoting proliferation [13]. Using engineered 3D tumor organoids with integrin-specific adhesive peptides, they observed increased proliferation, clustering, and resistance to both chemotherapies and the histone deacetylase inhibitor, panobinostat.

4.7 Applications of Lymphoma Organoid Models in Immunotherapy Research

Lymphoma treatment has evolved significantly with the introduction of immunotherapy [21, 38]. These therapies include monoclonal antibodies, immune checkpoint inhibitors, and CAR T cell therapies. While these treatments have shown remarkable success, a deeper understanding of TME, specifically the complex interactions between tumor cells, immune cells, cytokines, and suppressive factors, is critical to expanding the benefits of immunotherapy to a larger patient population [20]. Lymphoma organoid models serve as valuable platforms for detailed investigations into the TME, allowing for the exploration of resistance mechanisms to immunotherapies and the development of novel therapeutic strategies by targeting newly discovered signaling pathways and intercellular interactions. Kastenschmidt et al. used FL PDLOs to evaluate the effects of CD3:CD19 and CD3:CD20 bispecific antibodies, demonstrating specific B cell elimination and T cell activation [8]. Their findings suggest a complex role for T follicular helper (T_{fh}) cells: a higher initial T_{fh} presence in the TME was associated with poorer outcomes, while their activation post-treatment improved tumor cell elimination. The authors proposed that appropriately activated T_{fh} cells may enhance the efficacy of bispecific antibody therapy.

In addition to modeling individual drug responses, PDLOs allow for the assessment of combinatorial treatment strategies [10]. For example, given the frequent co-occurrence of molecular alterations in the histone methyltransferase EZH2 and the antiapoptotic protein Bcl-2 in DLBCL, Scholze et al. investigated the synergistic potential of dual EZH2 and Bcl-2 inhibition [39]. They demonstrated that the combination of tazemetostat (an EZH2 inhibitor) and venetoclax (a Bcl-2 inhibitor) resulted in enhanced anti-lymphoma activity in EZH2-mutant, BCL2-rearranged DLBCL models, including lymphoma organoids and patient-derived xenografts (PDXs). While immunotherapeutic agents were not used in this study, it highlights the emerging utility of lymphoma organoids as preclinical platforms for evaluating novel therapeutic strategies, including combinations of targeted therapies and immunotherapies.

4.8 Case Studies—Organoid Models Predicting Patient Immunotherapy Responses

Organoid-based immunotherapy assays have begun to show promising concordance with actual patient outcomes, laying the groundwork for personalized predictive testing. Several high-profile case studies demonstrate the efficacy of organoid models in forecasting how a patient's cancer will respond to immunotherapies:

- *Follicular Lymphoma and Bispecific T Cell Engagers*: Kastenschmidt et al. recently developed PDLOs from FL lymph node biopsies, which maintain the autologous tumor B cells together with their native T cell microenvironment [8] . When these PDLOs were treated with CD3:CD19 or CD3:CD20 bispecific antibodies, they exhibited patient-specific response profiles that mirrored known clinical heterogeneity to these agents. In the ex vivo model, CD3:CD19 engagers induced robust lymphoma cell killing in all organoids, whereas CD3:CD20 engagers showed variable, attenuated efficacy. By interrogating the PDLOs, the study identified correlates of response—for example, greater activation of $CD4^{+}$ T cells (particularly T_{fh} cells) within the organoid was associated with more efficient tumor cell elimination. These findings suggest that a patient's organoid could potentially predict which immunotherapeutic (if any) will maximally activate their tumor-infiltrating T cells. While in this study the organoid results were not yet used to direct therapy, they retrospectively mirrored the expectation that CD3:19 bispecific (like blinatumomab) are very potent in B cell malignancies, whereas other engagers can be less effective—and critically, the platform allows *quantification* of that efficacy for each patient in vitro.
- *Mantle Cell Lymphoma (MCL) and Combined Therapy Responses:* Araujo-Ayala et al. established a 3D organoid model for MCL that incorporated not only the malignant B cells but also supportive stromal and allogeneic myeloid cells to mimic the lymph node niche [18]. This complex co-culture recapitulated the lymphoma's native signaling and immune profile, enabling accurate prediction of therapeutic responses. Strikingly, the patient-derived MCL organoids responded to ibrutinib in vitro in concordance with the patients' known in vivo responses. Sensitive cases showed organoid growth inhibition and apoptosis

upon drug exposure, whereas organoids from ibrutinib-resistant patients did not. Moreover, the presence of an immune compartment in these "lymphoma organoids" allowed assessment of immune checkpoint blockade: when an anti-PD-1 antibody was added, some organoids exhibited enhanced T cell activation and tumor cell killing ex vivo, corresponding to cases with inflamed, immune-responsive tumor microenvironments. This case exemplifies how integrating immune context in organoids permits not only replication of targeted therapy efficacy, but also combined or sequential treatments (e.g., ibrutinib plus ICI) to identify synergistic or antagonistic effects relevant to patient care.

4.9 Limitations of Lymphoma Organoid Models

Despite their promise, lymphoma organoid models face significant scientific and translational challenges. Lymphomas are inherently complex hematologic malignancies, and recreating their TME in vitro is more difficult than for solid epithelial tumors [10, 20].

- *Establishment and Viability of Lymphoma Cells:* Primary lymphoma cells often have strict growth requirements, relying on supportive stromal cells and cytokines from their native lymphoid niche [20]. Outside the body, they tend to undergo apoptosis [10]. Early lymphoma co-culture attempts often depended on feeder cell lines or exogenous growth factors to sustain malignant B cells [9, 10]. The recent FL PDLO platform addressed viability by incorporating the *entire* cell population from the patient's lymph node, thereby preserving crucial cell–cell interactions and obviating the need for external growth factors [8, 16]. This allowed malignant B cells to survive in vitro for 2–3 weeks.
- *Immune Cell Longevity and Function:* Lymphoma organoids rely on *co-cultured immune cells* to model immunotherapy, but maintaining functional immune cells ex vivo poses another challenge. T cells in culture can become exhausted, undergo activation-induced cell death, or simply lose effector function over time without continuous stimulation [34]. In PDLO models, for instance, T cells gradually differentiated during the 21-day culture [8]. By the end of the culture, the T cell compartment may no longer mirror the fresh tumor. This limits the window for testing therapies. Moreover, if an organoid lacks certain immune subset (e.g., dendritic cells or neutrophils), it cannot fully replicate immune interactions [27, 34]. Most current lymphoma organoids focus on T cells and B cells, but omit other marrow- or lymph node-derived components [10]. This *limited immune repertoire* means an organoid might respond to an immunotherapy differently than an intact immune system would [10, 27]. Researchers are beginning to add these components (e.g., supplementing organoids with autologous monocyte-derived dendritic cells or macrophages), but standard methods to do so are not yet established [16, 18, 34].

- *Microenvironment and Structural Fidelity:* Lymphoid organs have a complex 3D architecture and vasculature that orchestrate immune cell trafficking [20]. However, organoids lack blood vessels and lymphatic drainage. Consequently, they cannot model the dynamic influx of immune cells or drugs from circulation [10, 27]. The absence of vascular endothelial cells and flow also affects gradients of oxygen, nutrients, and waste [40]. Furthermore, without vasculature, issues like *immune cell extravasation* or homing to tumor sites cannot be simulated [40]. This limitation is being tackled by organ-on-a-chip technologies: for example, microfluidic devices have been designed as "lymph node-on-a-chip" to impose shear flow and examine how T cells physically interact with dendritic cells under flow conditions [41]. Such bioengineering advances may eventually be combined with organoids to incorporate pseudo-vascular networks and fluid flow, but these remain complex experimental setups, not yet routine for personalized testing.
- *Inter-patient Variability and Standardization:* Lymphomas are highly heterogeneous between patients. Organoid models faithfully reflect this individuality, which is valuable for personalized medicine but problematic for generalizable conclusions. Reproducibility on a large scale will require rigorous protocol optimization for each lymphoma subtype. Additionally, the success rate of organoid establishment varies by lymphoma subtype and disease state—indolent lymphomas from lymph nodes might form organoids more readily than aggressive leukemic phase lymphomas or post-therapy samples [8, 10, 22]. This selection bias could limit utility if, for example, only untreated patients' samples yield organoids, whereas one would most need the model for refractory disease cases.
- *Translational and Practical Hurdles:* Even if a lymphoma organoid is successfully grown and tested with an array of immunotherapies, translating those results to patient care involves practical considerations. Therapy often cannot wait more than a couple of weeks; thus, organoid testing must be streamlined. Another issue is regulatory acceptance—before clinicians rely on organoid predictions, the approach requires validation in clinical trials and likely regulatory approval as a diagnostic test. Ensuring quality control (that the organoid truly represents the patient's current disease) is essential. Regular molecular profiling of organoids compared to original tumor can help monitor this, but adds cost and complexity. Organoid culture requires specialized reagents (e.g., growth factors, 3D matrices) and skilled personnel, meaning widespread adoption would incur significant expense. Finally, certain immunotherapeutic mechanisms are hard to emulate: for example, monoclonal antibodies that rely on complement or Fc-mediated opsonization might not function in minimal media without serum or with absent immune effector subsets. A classic example is rituximab for B cell lymphoma—its efficacy in patients involves NK cells and macrophages, which need to be present in the organoid system to see a meaningful effect [18, 27, 34]. If an organoid lacks sufficient FcγR-bearing cells, one might falsely conclude rituximab has no activity.
- *Hurdles in Fully Recapitulating the In Vivo TME:* B cell lymphomas are highly dependent on their surrounding niche. However, standard organoid culture

methods often strip away these non-malignant components, resulting in simplified models that lack critical microenvironmental features [10, 27]. In particular, it remains a major challenge to model an intact lymphoma TME with native vasculature, stromal support, and immune infiltrates in vitro [10, 24]. Without perfused vasculature, organoids rely on passive diffusion for oxygen and nutrients, leading to unnaturally uniform conditions or the development of necrotic cores only when organoids grow beyond a certain size [40]. They do not experience the gradients in oxygen tension or acidic pH that characterize many tumors, nor do they recreate the endothelial barrier that immune cells must traverse to enter tumor sites [40]. This is a fundamental limitation because the presence of vessels in the TME influences cancer cell metabolism, gene expression, and therapeutic responses.

- *Lack of Stromal Support:* Solid and hematologic tumors thrive within a supportive scaffold of stromal cells and ECM. In lymphomas, the stromal compartment includes fibroblastic reticular cells and follicular dendritic cells that organize lymphoid structure and provide survival signals to malignant B cells [1, 20]. Yet typical PDO protocols enrich for the tumor cells while largely eliminating stromal cells and ECM [10, 27]. Without cancer-associated fibroblasts (CAFs) or native ECM, the organoid cannot replicate tumor–stroma crosstalk such as growth factor exchange, cytokine signaling, and matrix remodeling [1, 10, 20]. These interactions are known to modulate tumor behavior. Their absence represents an intrinsic limitation of traditional organoid systems.
- *Inadequate Immune–Tumor Interaction Fidelity:* Lymphomas are prototypically immune-related cancers. Ex vivo organoid cultures, however, lose the tumor-infiltrating immune cells present in the original tumor during tissue dissociation and culture establishment [10, 27]. This deprives the model of endogenous cytotoxic T cells, helper T cells, regulatory T cells, macrophages, and other leukocytes that normally surround and influence the malignant cells [34]. Even when some immune cells are initially retained, they often survive only short term [8, 9]. The native immune microenvironment provides both anti-tumor surveillance and pro-tumor support (through exhausted T cells, M2-polarized macrophages, etc.), and this balance is often perturbed in cancer [34]. Organoid models lacking functional immune infiltrates cannot capture phenomena like immune evasion or the mechanisms of action of immunotherapies [27, 34]. They also cannot account for patient-specific immune responses, which are especially pertinent in lymphomas where therapies like CAR T cells or bispecific antibodies enlist the immune system [21, 27].
- *Simplified Spatial Architecture and Heterogeneity:* In an intact lymphoma, malignant cells and various microenvironmental cells are organized in a complex 3D architecture, often with discernible substructures (e.g., light and dark zones in FL germinal centers, or tumor cell-rich vs. T cell-rich regions) [1, 20]. This spatial heterogeneity creates gradients in signals and local cell–cell interactions that are critical to tumor biology. Organoids, however, lack the higher-order tissue architecture of lymphoid organs. When tumor cells are grown as free-floating spheroids or in a homogeneous gel, the fine spatial organization is lost. Immune

and stromal cells, if present at all, are not arranged as in a lymph node follicle; instead, they may be randomly intermixed. Traditional organoids do not recapitulate native tissue topology or compartmentalization, so they may fail to reproduce location-dependent phenomena [27]. Recent studies emphasize that spatial distribution of cells in the TME can affect therapeutic outcomes [28, 32]. Without spatial heterogeneity, organoids present an oversimplified tumor ecosystem, which might overestimate treatment efficacy or miss microenvironment-mediated resistance that only arises in certain tumor niches.

4.10 Strategies to Overcome Current Limitations and Improve Model Fidelity

In this section, we highlight several emerging strategies—including microfluidic systems (organoids-on-chips), 3D bioprinting, refined matrix design, and patient-specific co-cultures—that aim to overcome the technical limitations outlined above. These approaches are not mutually exclusive; in fact, the convergence of technologies appears especially promising for constructing next-generation lymphoma models. The overarching goal of all these strategies is to increase the fidelity of organoids to the native tumor, improving their utility for preclinical drug testing and personalized medicine.

- *Microfluidic Organoid-on-Chip Platforms:* Organoid-on-chip technology provides a more lifelike microenvironment with perfused vasculature and mechanical forces, directly tackling two major shortcomings of conventional organoids. While these systems require specialized equipment and expertise in microfabrication, they are becoming more accessible, and their design is highly modular—multiple traps or wells can be incorporated to run parallel organoid experiments under flow. For lymphoma modeling, a "lymphoma-on-chip" could maintain a miniaturized circulation to study how immune cells or therapeutics migrate into tumor spheroids, or how fluid dynamics influence the distribution of nutrients and signaling molecules in a pseudo-lymph node [24, 41]. Integrating microfluidics is therefore a key step toward recreating the vascularized and dynamic nature of the TME in organoids.
- *3D Bioprinting of Organoids and Tissues:* In bioprinting, living cells and bioinks (hydrogel matrices with biochemical cues) are deposited layer-by-layer according to a pre-designed architecture [42]. This enables precise control over the spatial placement of multiple cell types and matrix components in a tissue construct. The ability of bioprinting to position diverse cell types in defined 3D arrangements addresses the limitation of spatial heterogeneity [24, 42]. Moreover, bioprinting offers high reproducibility in organoid size and geometry, overcoming variability in shape/size [26, 42]. While still an emerging technology in the cancer modeling field, bioprinting significantly expands the complexity and

scale of organoid models, making it a promising tool to engineer more faithful replicas of the lymphoma microenvironment.

- *Refinement of Extracellular Matrix and Culture Scaffolds:* A subtler but equally important strategy for improving organoid fidelity lies in optimizing the ECM context in which organoids grow. For lymphomas, a decellularized lymphoid tissue matrix might provide the adhesion molecules and cytokine reservoirs that lymphoma cells encounter in lymph nodes [10, 43]. Another approach is synthetic hydrogels with tunable stiffness and degradability [4, 13, 24]. By adjusting polymer composition and crosslinking, these matrices can be made to match the mechanical stiffness of lymphoid tissue or to dynamically soften/stiffen in response to cell remodeling. In essence, refining the culture matrix addresses both the biochemical and biomechanical voids in standard organoid cultures. By providing the right integrin ligands, matrix-bound growth factors, and mechanical resistance, these advanced matrices encourage organoid-grown lymphoma cells to behave as they would in their natural habitat. They can also enhance the longevity and stability of co-cultured immune cells or stroma [13, 18, 43].
- *Patient-Specific Multicell-Type Co-cultures:* Finally, a transformative strategy to improve model fidelity is the incorporation of patient-derived supporting cells—immune and stromal—alongside the tumor cells in organoid cultures. One way to achieve this is by starting with intact tumor tissue fragments in culture. This approach has been demonstrated in FL: researchers have established PDLOs and spheroids that preserve the endogenous T cells and macrophages from the original tumor biopsy [7, 8, 16]. Remarkably, these lymphoma organoids maintained a stable cellular composition over multiple weeks in culture. The ability to sustain autologous immune cells and stromal elements for extended periods directly addresses the short-term viability issue in standard cultures [8, 16, 18]. Another approach to patient-specific co-culture is to add back patient immune cells to established tumor organoids [18, 27]. For example, autologous peripheral blood lymphocytes can be introduced into a tumor organoid culture to recreate tumor–immune interactions [18, 27]. Co-cultures can also include autologous mesenchymal stromal cells from the patient's bone marrow or lymph node, which may improve tumor cell viability and drug resistance in culture, emulating protective niches [16, 18, 27]. The key advantage of patient-specific co-cultures is the retention of native cell–cell interactions and the ability to observe therapeutic effects on all cell types. These complex co-cultures do come with practical challenges: they require sufficient tumor material to start, and the presence of many cell types can introduce variability and complicate readouts. Nevertheless, protocols have now been developed to reproducibly grow such multi-lineage organoid cultures [9, 16, 18].

4.11 Future Directions in Lymphoma Organoid Research

4.11.1 Emerging Technologies and Methodologies in Organoid Research

Modern lymphoma organoid research leverages a suite of innovative technologies to better recapitulate tumor biology and interrogate complex behaviors. Applied to patient-derived lymphoma cultures, spatial transcriptomic profiling can identify distinct cellular niches and confirm that ex vivo organoids preserve the heterogeneous architecture of the original tumor tissue [32]. Notably, in FL "lymphomoid" models, spatial multi-omics demonstrated maintenance of native B cell/immune cell organization and enabled assessment of drug effects on both tumor and microenvironment compartments [9, 32]. In parallel, machine learning algorithms now permit label-free, high-content analysis of organoid images, automatically segmenting cells, and even inferring fluorescent signals from bright-field microscopy [28, 33]. Together, spatial omics and AI-driven image analysis equip researchers to dissect lymphoma organoids with unprecedented resolution, ensuring that subtle phenotypic or molecular changes are detected and linked to functional outcomes.

High-throughput culture methods are being adapted for lymphoma organoids to enable large-scale drug and genetic screens. In an innovative approach, 3D bioprinting has been used to precisely pattern lymphoma cells into mini-organoid droplets, yielding uniform 3D microtissues with preserved histology and gene expression [26, 42]. These bioprinted organoids are compatible with automated handling and imaging, greatly enhancing throughput. Additionally, CRISPR/Cas9 genome editing is being integrated into organoid systems to probe lymphoma genetics in three dimensions [35]. CRISPR knockout screens performed in cancer organoids have revealed 3D-specific vulnerabilities—genes essential for growth in a 3D context but not in 2D cultures [10, 27, 35]. This underscores how lymphoma organoids can uncover novel drug targets missed by traditional models.

Another critical methodological frontier is the dynamic modeling of the TME within lymphoma organoids. Recent PDLOs maintain a rich cellular ecosystem, including proliferating malignant B cells co-mingled with patient-derived T cells and macrophages, faithfully reproducing germinal center-like interactions [8, 16, 18]. Remarkably, these organoids preserve T cell subsets (e.g., T_{fh} and T_{reg} cells) and even the spectrum of T cell activation/exhaustion phenotypes seen in the patient's tumor. The ability to sustain syngeneic tumor-infiltrating lymphocytes for weeks is a breakthrough for immunologic modeling [34]. It enables lymphoma organoids to recapitulate immune evasion mechanisms and antibody-dependent cell cytotoxicity in vitro. Such advances in TME co-culture, along with the above technologies, position lymphoma organoids as increasingly faithful "micro-tumors" where spatial context, cellular diversity, and real-time dynamics of lymphoma can all be interrogated.

4.11.2 Potential for Personalized Medicine Applications

Lymphoma organoids are poised to become invaluable tools in precision oncology, enabling patient-specific therapy optimization through ex vivo testing. One promising application is drug sensitivity profiling using PDOs to predict which therapies a given lymphoma will respond to. Studies now indicate that responses observed in lymphoma organoids often correlate with clinical outcomes [9, 16, 18]. As organoid biobanks and screening pipelines expand, this approach could spare patients from trial-and-error treatments by identifying responsive versus resistant therapies *before* they are given, truly personalizing lymphoma care. In addition to cytotoxic or targeted drugs, lymphoma organoids offer a testing ground for personalized immunotherapy selection. Because cutting-edge organoid models preserve the patient's immune microenvironment, they can be used to evaluate immune-based treatments in a patient-specific context [34]. By capturing the intricacies of individual tumor biology—drug sensitivities, resistance pathways, and immune interactions—lymphoma organoids provide a personalized testing platform that can refine therapy choices and improve outcome predictions on a case-by-case basis.

4.12 Conclusion

In summary, lymphoma organoid models have emerged as transformative ex vivo platforms that faithfully recapitulate the TME. By establishing patient-derived three-dimensional cultures that preserve the native architecture and cellular heterogeneity of lymphoid malignancies, these models address a longstanding gap left by traditional cell lines and xenografts. Unlike earlier lymphoma spheroids or aggregates that lacked the complex spatial organization of real tissue, modern organoids—such as FL patient-derived spheroids—maintain malignant B cells alongside autologous immune constituents in 3D arrangements that mirror lymph node architecture. This fidelity to the in vivo tumor milieu enables more accurate modeling of tumor–immune interactions, which is critical given that immune evasion and microenvironmental signals heavily influence lymphoma progression and therapeutic response. Notably, investigators can now sustain these miniaturized lymphomas in culture for multiple weeks without exogenous growth factors, underscoring the robustness of the organoid system and establishing a patient-specific platform for interrogating lymphoma biology. Looking ahead, the convergence of immunology, bioengineering, and genomics promises to further enhance lymphoma organoid models. Future efforts are focusing on incorporating missing cellular constituents and extending culture longevity, thereby creating more physiologically complete "lymphoma organoids" that can capture tumor–immune dynamics over extended periods. Innovative interdisciplinary approaches—including microfluidic organ-on-a-chip systems, 3D bioprinting, and high-resolution single-cell analytics—are being applied to simulate aspects like vascular perfusion, immune cell trafficking, and clonal evolution within organoids. These technological enhancements will not only improve the fidelity of the model but also enable high-throughput drug and cell therapy screening under conditions that closely approximate a patient's lymph node microenvironment. In conclusion, lymphoma organoid models illustrate the profound potential of technological convergence in oncology, uniting advances in 3D culture, immune engineering, and personalized genomics to transform how we study and ultimately treat lymphoma.

References

1. Ame-Thomas P, Tarte K. The yin and the yang of follicular lymphoma cell niches: role of microenvironment heterogeneity and plasticity. Semin Cancer Biol. 2014;24:23–32. Epub 2013/08/28.
2. Bhandari V, Li CH, Bristow RG, Boutros PC, Consortium P. Divergent mutational processes distinguish hypoxic and normoxic tumours. Nat Commun. 2020;11(1):737. Epub 2020/02/07.
3. Alizadeh AA, Gentles AJ, Alencar AJ, Liu CL, Kohrt HE, Houot R, et al. Prediction of survival in diffuse large B-cell lymphoma based on the expression of 2 genes reflecting tumor and microenvironment. Blood. 2011;118(5):1350–8. Epub 2011/06/15.
4. Chrisnandy A, Blondel D, Rezakhani S, Broguiere N, Lutolf MP. Synthetic dynamic hydrogels promote degradation-independent in vitro organogenesis. Nat Mater. 2022;21(4):479–87. Epub 2021/11/17.
5. Bar-Ephraim YE, Kretzschmar K, Clevers H. Organoids in immunological research. Nat Rev Immunol. 2020;20(5):279–93. Epub 2019/12/20.
6. Britto LS, Balasubramani D, Desai S, Phillips P, Trehan N, Cesarman E, et al. T cells spatially regulate B cell receptor signaling in lymphomas through H3K9me3 modifications. Adv Healthc Mater. 2025;14(5):e2401192. Epub 2024/06/05.
7. Dobano-Lopez C, Valero JG, Araujo-Ayala F, Nadeu F, Gava F, Faria C, et al. Patient-derived follicular lymphoma spheroids recapitulate lymph node signaling and immune profile uncovering galectin-9 as a novel immunotherapeutic target. Blood Cancer J. 2024;14(1):75. Epub 2024/05/03.
8. Kastenschmidt JM, Schroers-Martin JG, Sworder BJ, Sureshchandra S, Khodadoust MS, Liu CL, et al. A human lymphoma organoid model for evaluating and targeting the follicular lymphoma tumor immune microenvironment. Cell Stem Cell. 2024;31(3):410–20.e4. Epub 2024/02/25.
9. Santamaria-Martinez A, Epiney J, Srivastava D, Tavernari D, Varrone M, Milowich D, et al. Development of patient-derived lymphomoids with preserved tumor architecture for lymphoma therapy screening. Nat Commun. 2024;15(1):10650. Epub 2024/12/10.
10. Mattioda C, Voena C, Ciardelli G, Mattu C. In vitro 3D models of haematological malignancies: current trends and the road ahead? Cells. 2025;14(1):38. Epub 2025/01/10.
11. Sabhachandani P, Sarkar S, McKenney S, Ravi D, Evens AM, Konry T. Microfluidic assembly of hydrogel-based immunogenic tumor spheroids for evaluation of anticancer therapies and biomarker release. J Control Release. 2019;295:21–30. Epub 2018/12/15.
12. Foxall R, Narang P, Glaysher B, Hub E, Teal E, Coles MC, et al. Developing a 3D B cell lymphoma culture system to model antibody therapy. Front Immunol. 2020;11:605231. Epub 2021/02/26.
13. Tian YF, Ahn H, Schneider RS, Yang SN, Roman-Gonzalez L, Melnick AM, et al. Integrin-specific hydrogels as adaptable tumor organoids for malignant B and T cells. Biomaterials. 2015;73:110–9. Epub 2015/09/26.

14. Saba NS, Liu D, Herman SE, Underbayev C, Tian X, Behrend D, et al. Pathogenic role of B-cell receptor signaling and canonical NF-kappaB activation in mantle cell lymphoma. Blood. 2016;128(1):82–92. Epub 2016/04/30.
15. Bahlmann LC, Xue C, Chin AA, Skirzynska A, Lu J, Theriault B, et al. Targeting tumour-associated macrophages in hodgkin lymphoma using engineered extracellular matrix-mimicking cryogels. Biomaterials. 2023;297:122121. Epub 2023/04/20.
16. Faria C, Gava F, Gravelle P, Valero JG, Dobano-Lopez C, Van Acker N, et al. Patient-derived lymphoma spheroids integrating immune tumor microenvironment as preclinical follicular lymphoma models for personalized medicine. J Immunother Cancer. 2023;11(10):e007156. Epub 2023/10/30.
17. Aldinucci D, Celegato M, Casagrande N. Microenvironmental interactions in classical Hodgkin lymphoma and their role in promoting tumor growth, immune escape and drug resistance. Cancer Lett. 2016;380(1):243–52. Epub 2015/10/18.
18. Araujo-Ayala F, Dobano-Lopez C, Valero JG, Nadeu F, Gava F, Faria C, et al. A novel patient-derived 3D model recapitulates mantle cell lymphoma lymph node signaling, immune profile and in vivo ibrutinib responses. Leukemia. 2023;37(6):1311–23. Epub 2023/04/09.
19. Casagrande N, Borghese C, Aldinucci D. Current and emerging approaches to study microenvironmental interactions and drug activity in classical Hodgkin lymphoma. Cancers. 2022;14(10):2427. Epub 2022/05/29.
20. Scott DW, Gascoyne RD. The tumour microenvironment in B cell lymphomas. Nat Rev Cancer. 2014;14(8):517–34. Epub 2014/07/11.
21. Schuster SJ, Svoboda J, Chong EA, Nasta SD, Mato AR, Anak O, et al. Chimeric antigen receptor T cells in refractory B-cell lymphomas. N Engl J Med. 2017;377(26):2545–54. Epub 2017/12/12.
22. Yanguas-Casas N, Pedrosa L, Fernandez-Miranda I, Sanchez-Beato M. An overview on diffuse large B-cell lymphoma models: towards a functional genomics approach. Cancers. 2021;13(12):2893. Epub 2021/07/03.
23. Neal JT, Li X, Zhu J, Giangarra V, Grzeskowiak CL, Ju J, et al. Organoid modeling of the tumor immune microenvironment. Cell. 2018;175(7):1972–88.e16. Epub 2018/12/15.
24. Wang X, Luo Y, Ma Y, Wang P, Yao R. Converging bioprinting and organoids to better recapitulate the tumor microenvironment. Trends Biotechnol. 2024;42(5):648–63. Epub 2023/12/10.
25. Aljitawi OS, Li D, Xiao Y, Zhang D, Ramachandran K, Stehno-Bittel L, et al. A novel three-dimensional stromal-based model for in vitro chemotherapy sensitivity testing of leukemia cells. Leuk Lymphoma. 2014;55(2):378–91. Epub 2013/04/10.
26. Tebon PJ, Wang B, Markowitz AL, Davarifar A, Tsai BL, Krawczuk P, et al. Drug screening at single-organoid resolution via bioprinting and interferometry. Nat Commun. 2023;14(1):3168. Epub 2023/06/07.
27. Xu H, Jiao D, Liu A, Wu K. Tumor organoids: applications in cancer modeling and potentials in precision medicine. J Hematol Oncol. 2022;15(1):58. Epub 2022/05/14.
28. Yan H, Ju X, Huang A, Yuan J. Advancements in technology for characterizing the tumor immune microenvironment. Int J Biol Sci. 2024;20(6):2151–67. Epub 2024/04/15.
29. Katt ME, Placone AL, Wong AD, Xu ZS, Searson PC. In vitro tumor models: advantages, disadvantages, variables, and selecting the right platform. Front Bioeng Biotechnol. 2016;4:12. Epub 2016/02/24.
30. Porter RJ, Murray GI, McLean MH. Current concepts in tumour-derived organoids. Br J Cancer. 2020;123(8):1209–18. Epub 2020/07/31.
31. Gu Z, Wu Q, Shang B, Zhang K, Zhang W. Organoid co-culture models of the tumor microenvironment promote precision medicine. Cancer Innov. 2024;3(1):e101. Epub 2024/07/01.
32. Pickard K, Stephenson E, Mitchell A, Jardine L, Bacon CM. Location, location, location: mapping the lymphoma tumor microenvironment using spatial transcriptomics. Front Oncol. 2023;13:1258245. Epub 2023/10/23.

33. Shi H, Kowalczewski A, Vu D, Liu X, Salekin A, Yang H, et al. Organoid intelligence: integration of organoid technology and artificial intelligence in the new era of in vitro models. Med Novel Technol Devices. 2024;21:100276. Epub 2024/04/22.
34. Sun CP, Lan HR, Fang XL, Yang XY, Jin KT. Organoid models for precision cancer immunotherapy. Front Immunol. 2022;13:770465. Epub 2022/04/23.
35. Han K, Pierce SE, Li A, Spees K, Anderson GR, Seoane JA, et al. CRISPR screens in cancer spheroids identify 3D growth-specific vulnerabilities. Nature. 2020;580(7801):136–41. Epub 2020/04/03.
36. Li S, Ren J, Wu J, Xia Z, Li Y, Li C, et al. Establishment and molecular characterisation of patient-derived organoids for primary central nervous system lymphoma. Leukemia. 2025;39(5):1169–83. Epub 2025/03/19.
37. Shah SB, Carlson CR, Lai K, Zhong Z, Marsico G, Lee KM, et al. Combinatorial treatment rescues tumour-microenvironment-mediated attenuation of MALT1 inhibitors in B-cell lymphomas. Nat Mater. 2023;22(4):511–23. Epub 2023/03/18.
38. Pizzi M, Boi M, Bertoni F, Inghirami G. Emerging therapies provide new opportunities to reshape the multifaceted interactions between the immune system and lymphoma cells. Leukemia. 2016;30(9):1805–15. Epub 2016/07/09.
39. Scholze H, Stephenson RE, Reynolds R, Shah S, Puri R, Butler SD, et al. Combined EZH2 and Bcl-2 inhibitors as precision therapy for genetically defined DLBCL subtypes. Blood Adv. 2020;4(20):5226–31. Epub 2020/10/27.
40. Kim JW, Ho WJ, Wu BM. The role of the 3D environment in hypoxia-induced drug and apoptosis resistance. Anticancer Res. 2011;31(10):3237–45. Epub 2011/10/04.
41. Moura Rosa P, Gopalakrishnan N, Ibrahim H, Haug M, Halaas O. The intercell dynamics of T cells and dendritic cells in a lymph node-on-a-chip flow device. Lab Chip. 2016;16(19):3728–40. Epub 2016/08/26.
42. Datta P, Dey M, Ataie Z, Unutmaz D, Ozbolat IT. 3D bioprinting for reconstituting the cancer microenvironment. NPJ Precis Oncol. 2020;4:18. Epub 2020/08/15.
43. Xia T, Du WL, Chen XY, Zhang YN. Organoid models of the tumor microenvironment and their applications. J Cell Mol Med. 2021;25(13):5829–41. Epub 2021/05/26.

Chapter 5
The Use of Organoids for Modeling Bone Tumors and Testing Drugs

Doruk Somuncu, Sefik Murat Arikan, and Ozge Sezin Somuncu

Abstract Bone tumor organoids (BTOs) are three-dimensional culture systems developed from patient-derived tissues, which accurately replicate the growth dynamics and cellular interactions characteristic of bone tumors in vitro. Organoids offer improved physiological relevance compared to traditional models, thereby enhancing the precision of therapeutic strategies and addressing existing treatment limitations. The effectiveness of current bone tumor therapies is often limited by issues such as drug resistance, recurrence, and metastasis. While still in early stages, organoid technology has shown promise in studying various types of bone tumors, including primary, metastatic, and bone marrow–origin tumors. Here we outline the methodologies for establishing bone tumor organoids, highlight their current and potential applications across different bone tumor subtypes, and examine their integration with novel technologies. Furthermore, this review discusses existing challenges and future research directions.

Keywords Bone tumors · Patient-derived tumor organoids (PDTOs) · Bone tumor organoids (BDTOs) · Personalized medicine · Translational research

D. Somuncu (✉) · S. M. Arikan
Gazi University, Department of Orthopedics and Traumatology, Ankara, Turkey
e-mail: smuratarikan@gazi.edu.tr

O. S. Somuncu
Center for DNA Damage and Repair, Dana-Farber Cancer Institute, Boston, MA, USA

Department of Radiation Oncology, Dana-Farber/Harvard Cancer Center, Boston, MA, USA
e-mail: ozges_somuncu@dfci.harvard.edu

O. S. Somuncu (ed.), *Organoids for Preclinical Disease Modeling and Personalized Medicine*, Stem Cell Biology and Regenerative Medicine 78,
https://doi.org/10.1007/978-3-032-10024-5_5

5.1 Introduction

Bone tumors are generally classified as either benign or malignant, with malignant forms representing a major clinical challenge [1]. These malignant bone tumors are further divided into primary and metastatic categories. Primary bone tumors, including osteosarcoma (OS), Ewing sarcoma (ES), and chondrosarcoma (CS), typically occur in younger individuals and are noted for their aggressive behavior and high metastatic potential [2]. In contrast, metastatic bone tumors commonly arise in patients with advanced cancers, particularly those with breast, prostate, or lung cancer [3].

Although conventional models have contributed significantly to bone tumor research, their intrinsic limitations impede the advancement of new therapeutic strategies. For example, while animal models can mimic certain aspects of tumor growth and metastasis, species-specific differences often result in poor translation of preclinical findings to clinical success. Similarly, two-dimensional (2D) cell culture systems fall short due to their lack of tissue architecture and tumor microenvironment (TME), as they involve uniform cell populations growing on flat, artificial surfaces. As a result, these models are unable to reflect the intrinsic heterogeneity of tumor cells [4].

Almost 20 years ago, Sung et al. were the first to establish a three-dimensional co-culture system combining human MG-63 or HS27A cells with C4-2 cells. Compared to conventional culture dishes, this 3D co-culture approach more effectively induced and maintained morphological changes in bone matrix cells (MG-63 or HS27A) [5]. Recent advances in bone tumor research have led to the development of three-dimensional tumor microenvironments (TMEs) using both synthetic and natural scaffolds, which more accurately mimic bone tumor growth dynamics, cell–cell interactions, and matrix associations under physiological conditions. Osteosarcoma (OS) organoids have been successfully generated as preclinical models, enabling high-throughput drug screening and the advancement of personalized therapies through single-cell sequencing techniques [6].

5.2 Producing Bone Tumor Organoids

Presently, research on bone tumor organoids remains in the exploratory stage, though various subtypes have already been successfully established.

5.2.1 Organoids Production from Different Patient Tissue Sources

Bone tumor organoids (BTOs) generated from biopsy samples begin with mechanical dissociation and enzymatic digestion, producing small tissue fragments. When cultured in a three-dimensional environment supplemented with specific growth factors, these cells self-organize and partially recapitulate the structural and functional features of the original tumor tissue. The resulting organoids mainly preserve patient-specific characteristics such as gene expression patterns, genetic mutations, and cellular morphology [7].

For metastatic bone tumors, tissue is generally obtained from the site of metastasis rather than the primary tumor, as metastatic site samples more accurately reflect the interactions between tumor cells and the bone microenvironment, rather than the characteristics of the primary tumor. Bone metastases from breast cancer (BCa BoM) are typically osteolytic and commonly occur in advanced stages of the disease. The limited availability of patient-derived (PD) samples is largely due to complications from chemotherapy, along with ethical and logistical difficulties in acquiring fresh tumor tissue [8]. In comparison, obtaining tissue samples from prostate cancer bone metastases (PCa BoM) is generally more feasible, as these lesions are predominantly osteoblastic and are typically detected earlier through imaging methods such as bone scans [9]. Lung cancer bone metastases (LCa BoM) primarily involve the spine, accounting for approximately 90% of cases, and are characterized by osteolytic lesions that result in vertebral degradation, often presenting with symptoms such as thoracic back pain and neurological deficits in the lower limbs [10]. Bone marrow-derived bone tumors, such as multiple myeloma (MM), engage in complex interactions with the surrounding bone marrow microenvironment, including stromal and immune cells. Due to the high degree of heterogeneity in these tumors, tissue samples alone are often insufficient to capture the full spectrum of tumor characteristics. Additionally, primary MM cells exhibit limited proliferative capacity under in vitro conditions. As a result, researchers frequently choose more easily manipulated cell lines for 3D culture studies [11].

Limitations associated with tissue sources include difficulties in sample acquisition, inherent tissue heterogeneity, challenges in maintaining cultures, and various ethical and legal considerations. Importantly, the central regions of bone tumors are often hypoxic and necrotic, whereas the peripheral regions tend to exhibit higher proliferative activity. Additionally, the collected tissue samples may reflect only certain areas of the bone tumor, potentially overlooking the tumor's regional heterogeneity and biological variability [12].

5.2.2 Techniques and Parameters for Organoid Cultures

Bone tumor organoids are typically cultured at 37 °C, in an atmosphere composed of 95% air and 5% CO_2, with 95%–100% humidity and a physiological pH range of 7.2–7.4. In certain cases, however, a hypoxic environment with 1%–5% oxygen and a reduced pH is used to better mimic the TME [13]. Mainly the base culture media used in studies are DMEM, RPMI-1640, or MEM, as they supply the fundamental nutrients required for cell growth. To enhance the proliferation of bone tumor cells, growth-promoting agents like epidermal growth factor (EGF), B-27 without vitamin A, fibroblast growth factor, and insulin-like growth factor are added.

Antioxidants like niacinamide and N-acetylcysteine are also incorporated to neutralize intracellular free radicals and protect cells from oxidative damage [2].

For primary bone tumors such as osteosarcoma (OS), chondromyxoid fibroma (CMA), and metastatic bone tumors, signaling pathway modulators are frequently introduced to control the balance between bone cell differentiation and proliferation. For instance, in the Wnt/β-catenin signaling pathway, commonly used modulators include RSPO1, the p38 MAPK inhibitor SB-202190, the ROCK inhibitor Y-27632, the TGF-β inhibitor A83-01, and BMP pathway inhibitors like Noggin [14–16]. Additionally, in the development of multiple myeloma (MM) organoids, the cytokine receptor activator of nuclear factor kappa-B ligand (RANKL) is introduced to promote osteoclast formation. Macrophage colony-stimulating factor (M-CSF) is also added, working in synergy with RANKL to enhance the bone resorption process [17, 18].

5.3 Research in Bone Tumor Organoids

Bone tumor organoid research offers a powerful platform for modeling tumor biology, enabling personalized drug testing, and advancing understanding of rare and aggressive bone cancers in a physiologically relevant 3D environment.

5.3.1 Osteosarcoma Organoids

Current research efforts have been largely concentrated on osteosarcoma (OS). Nie et al. successfully established OS organoids using 20 biopsy specimens and 12 surgical samples, achieving success rates of >90%, respectively. Remarkably, these organoids formed within 2 weeks and maintained stable phenotypic characteristics while proliferating over several months [16].

Another study has explored the therapeutic effects of pimozide, a STAT5 inhibitor that acts on the JAK/STAT signaling pathway [19]. By suppressing OS-related cell markers like DCLK1, CD44, and CD133, this approach reduces the growth of osteosarcoma stem cells and hinders the formation of 3D bone spheres [20]. Since current osteosarcoma treatments can be limited by resistance and relapse, this approach offers a promising alternative by targeting cancer stem cells and critical growth pathways.

Kinase inhibitors have been identified as promising therapeutic options for OS. However, because of the disease's complex pathogenesis, single-agent kinase inhibitors have shown limited clinical effectiveness. Recent research using OS organoids has demonstrated that testing different drug combinations revealed a synergistic effect between ALK and FLT3 inhibitors in suppressing OS growth [21]. While kinase inhibitors target specific enzymes important for cancer growth, using just one inhibitor typically has not worked well in clinical settings for OS, likely because tumors can bypass blocked pathways or activate alternative routes. The research with organoids particularly underlined synergistic effect of combining ALK and FLT3 kinase inhibitors.

Additionally, OS organoids offer a more accurate platform for studying how genetic changes like RB1 and TP53 deletions influence OS disease mechanisms. They also function as important preclinical tools for evaluating the anti-tumor effects of SKP2 inhibitors [21], and for confirming the expression profiles of PDGFD and PDGFRB, along with analyzing the spatial arrangement and binding properties of ligand-receptor interactions [22].

5.3.2 Chordoma Organoids

Due to its rarity, CMA poses considerable research difficulties, making patient-derived organoid models essential for gaining deeper insights into the disease's underlying mechanisms [23]. Scientists have effectively developed CMA organoids that closely replicate the tumor microenvironment (TME), maintaining essential immune elements such as CD8+ T cells and PD-L1–positive cells. This model has been utilized to test the effectiveness of the immunotherapy drugs like nivolumab, helping to speed up drug screening for this slow-progressing tumor [24].

B7–H3, a prominently expressed antigen on CMA cells, has emerged as a promising target for CAR-T cell therapy. Studies show that engineering B7-H3 CAR-7 cells to express IL-7 (B7-H3 CAR-7/IL-7) can effectively eliminate CMA cells in organoid models and demonstrate prolonged anti-tumor activity [25]. Demonstrating that these engineered CAR-T cells can eliminate CMA cells in organoid models is significant because organoids mimic the 3D structure and complexity of real tumors better than traditional cell cultures.

Shihabi and colleagues used CMA organoid models to efficiently screen hundreds of drug compounds, leading to the successful identification of several promising candidates, including PI3K/mTOR inhibitors like sapanisertib and vistusertib, the EGFR inhibitor gefitinib, and JAK2/STAT3 pathway inhibitors such as fedratinib [26]. This study was particularly important due to the pinpoints to several potential treatments like PI3K/mTOR inhibitors, an EGFR inhibitor, and JAK2/STAT3 pathway inhibitors, which might not have been obvious targets without this high-throughput screening.

5.3.3 Ewing Sarcoma Organoids

While tumors most frequently develop in children and young adults, individuals of any age can be affected by ES. It is traditionally associated with bone, but a substantial number of cases also arise in soft tissues, particularly among older patients [27].

Maurer et al. generated ES PDOs from a metastatic lung lesion in a patient carrying a hereditary BRCA1-associated RING domain 1 (BARD1) mutation. Their initial data demonstrated that the organoids showed a strong responsiveness to poly (ADP-ribose) polymerase (PARP) inhibitors [28]. Their second research demonstrated that BARD1 loss heightens Ewing sarcoma's sensitivity to DNA damage, with Guanylate-binding protein 1 contributing to the DNA damage response in these organoids. This finding provides valuable insight into key mechanisms driving Ewing sarcoma [29]. By connecting specific genetic mutations (like BARD1 loss) with drug sensitivity in a model that mimics the patient's tumor, this research moves us closer to precision oncology, where treatments are tailored based on the tumor's unique biology.

5.3.4 Chondrosarcoma Organoids

Chondrosarcomas (CS) are cancerous bone tumors characterized by a dense, cartilage-like extracellular matrix and a hypoxic microenvironment, both of which contribute to their resistance to chemotherapy and radiotherapy [30].

Veys and colleagues pioneered the use of chondrosarcoma spheroids derived from the SW1353 cell line to test the effects of microRNAs. They found that miR-342-5p promotes apoptosis, especially under hypoxic conditions, and to a lesser degree miR-491-5p induces cytotoxic effects, confirming Bcl-2 and Bcl-xL as direct targets in both normoxia and hypoxia [31]. Although this study was conducted with single-cell-derived spheroid systems, it can be considered as a useful study on chondrosarcoma using three-dimensional model systems. This was the first report using a 3D model in CS to recreate the tumor's hypoxic extracellular matrix and test miRNA efficacy in a biologically relevant context.

5.3.5 Malignant Giant Cell Tumor Organoids

Malignant giant cell tumors (MGCTs) are very uncommon bone sarcomas that develop from conventional giant cell tumors after extended treatment periods. Due to their rarity, there has been no detailed study of their molecular pathogenesis, and consequently, no standard treatment protocol has been defined [32].

Suzuki and colleagues focused on developing and genetically profiling a novel organoid model derived from a MGCT patient. The organoid was successfully established from patient tumor samples, maintaining the three-dimensional structure and cellular heterogeneity of the original tumor. Key genes such as CD44 and TWIST, known to promote aggressive tumor behavior and epithelial-mesenchymal transition (EMT), were highly expressed in the organoids of MGCT. Most importantly, they observed that the organoid model showed resistance to conventional chemotherapy agents, mirroring the clinical challenge of treating MGCT [32]. This study can be considered particularly important for closely mimicking the patient's tumor characteristics proving organoid models offer a more accurate representation of tumor behavior and response to treatments.

5.4 Mimicking the Tumor Microenvironment for Improved Drug Screening

By incorporating components such as cancer-associated fibroblasts, immune cells, extracellular matrix (ECM), and even vascular-like structures, bone tumor organoids can more accurately simulate in vivo conditions. This enhanced physiological relevance allows for more predictive drug screening, as it accounts for the influence of the TME on drug penetration, efficacy, and resistance mechanisms. As a result, using organoid-based models improves the translational value of preclinical testing and supports the identification of more effective, targeted therapies for bone cancer patients.

Specifically, advanced multicellular spheroid models of osteosarcoma have integrated tumor cells with mesenchymal stem cells (MSCs) and tumor-associated macrophages, successfully recreating M2-like immunosuppressive macrophage polarization and native ECM features seen in osteosarcoma [33]. Such models respond dynamically to immunomodulatory agents, shifting secretome profiles in ways that inform therapeutic targeting. This incorporation of TME elements enables organoids to better predict drug penetration, efficacy, and resistance mechanisms compared to simpler systems [33, 34].

A 2021 study introduced a PLMA-based hydrogel system, embedding osteosarcoma spheroids with human osteoblasts and MSCs. This platform supported tumor invasion, stromal interaction, and demonstrated increased doxorubicin resistance compared to scaffold-free spheroids—highlighting the role of biomimetic matrices in drug resistance modeling [35].

In summary, bone tumor organoids that faithfully replicate the tumor microenvironment—through incorporation of stromal, immune, and matrix components—offer a much more predictive and translationally relevant platform for preclinical drug testing in osteosarcoma and related bone malignancies as listed in Table 5.1.

5.5 Conclusion

Organoids have revolutionized the landscape of bone tumor research, offering an unprecedented window into the biology of some of the most complex and aggressive cancers. These miniature, three-dimensional replicas of patient tumors do what conventional models cannot—they capture the true essence of bone malignancies, from their intricate cellular diversity to the dense extracellular matrix and hostile microenvironments that define them.

Table 5.1 Bone tumor organoids and tumor microenvironment modeling

Study/authors	Tumor type	Model type	TME components included	Key outcomes/findings
Fior et al. (2024) (PubMed ID: 39505219) [33]	Osteosarcoma	3D multicellular tissue spheroids (MCTS)	Tumor cells, MSCs, M2-polarized macrophages	Mimicked pro-tumoral immune TME; model responded to immunomodulatory agents; supports immune-targeted drug screening
Pereira et al. (2021) (PubMed ID: 34303015) [35]	Osteosarcoma	PLMA hydrogel-based 3D model	Osteoblasts, MSCs, ECM mimic	Showed increased doxorubicin resistance vs. 2D; highlighted role of stromal support and matrix stiffness
Organotypic Tumor Slices (Canine/Murine Models) (PubMed ID: 34638373) [36]	Osteosarcoma	Ex vivo organotypic culture	Intact stromal and immune populations (TAMs, CAFs, CD45+ cells)	Maintained native architecture; revealed reduced chemo-sensitivity due to preserved microenvironment
He et al. (2020) (PubMed ID: 33079257) [37]	Metastatic osteosarcoma	Patient-derived lung metastasis organoids	Native tumor architecture and T cells preserved	Maintained morphology, Vimentin/Sox9 expression; suitable for immunotherapy and drug testing
Zhao et al. (2020) (PMID: 32668273) [38]	Osteosarcoma	3D Bioprinted scaffold + OS cells	Gelatin-based scaffold, tumor cells	Improved modeling of in vivo-like invasion and drug sensitivity patterns
Gai et al. (2025) (PMCID: PMC11937690) [39]	Bone organoids (regenerative focus)	DNA-CPO-ECM hybrid hydrogel	Decellularized ECM, mineralized phase, dynamic culture	Created vascularized bone organoids; relevant for future integration with tumor models

In rare cancers like OS, ES, and CS—where treatment options are limited and outcomes often grim—organoids stand as a beacon of hope. They empower researchers to decode the genetic chaos that drives these tumors, test therapies in real time, and envision a future where treatment is no longer based on trial and error, but on precise, personalized science.

Though the road ahead is not without obstacles due to scarce tissue samples, technical hurdles, and the complexity of precise modeling of the tumor microenvironment, the progress made is nothing short of extraordinary. As these models become more integrated into translational research and precision oncology, they hold significant potential to accelerate the discovery of effective therapies and improve clinical outcomes for patients with bone cancers.

References

1. Choi JH, Ro JY. The 2020 WHO classification of tumors of bone: an updated review. Adv Anat Pathol. 2021;28(3):119–38.
2. Ma Z, Liu Y, Chen R, Fan H, Kong L, Cao X. A novel perspective on bone tumors: advances in organoid research. Front Pharmacol. 2025;16:1550163.
3. Coleman RE, Croucher PI, Padhani AR, Clezardin P, Chow E, Fallon M, et al. Bone metastases. Nat Rev Dis Primers. 2020;6(1):83.
4. Duval K, Grover H, Han LH, Mou Y, Pegoraro AF, Fredberg J, et al. Modeling physiological events in 2D vs. 3D cell culture. Physiology (Bethesda). 2017;32(4):266–77.
5. Sung SY, Hsieh CL, Law A, Zhau HE, Pathak S, Multani AS, et al. Coevolution of prostate cancer and bone stroma in three-dimensional coculture: implications for cancer growth and metastasis. Cancer Res. 2008;68(23):9996–10003.
6. Huang B, Li G, Cao L, Wu S, Zhang Y, Li Z, et al. Nanoengineered 3D-printing scaffolds prepared by metal-coordination self-assembly for hyperthermia-catalytic osteosarcoma therapy and bone regeneration. J Colloid Interface Sci. 2024;672:724–35.
7. Yoshida GJ. Applications of patient-derived tumor xenograft models and tumor organoids. J Hematol Oncol. 2020;13(1):4.
8. Ding K, Chen F, Priedigkeit N, Brown DD, Weiss K, Watters R, et al. Single cell heterogeneity and evolution of breast cancer bone metastasis and organoids reveals therapeutic targets for precision medicine. Ann Oncol. 2022;33(10):1085–8.
9. Gao D, Vela I, Sboner A, Iaquinta PJ, Karthaus WR, Gopalan A, et al. Organoid cultures derived from patients with advanced prostate cancer. Cell. 2014;159(1):176–87.
10. Hu X, Huang W, Sun Z, Ye H, Man K, Wang Q, et al. Predictive factors, preventive implications, and personalized surgical strategies for bone metastasis from lung cancer: population-based approach with a comprehensive cancer center-based study. EPMA J. 2022;13(1):57–75.
11. Reagan MR, Mishima Y, Glavey SV, Zhang Y, Manier S, Lu ZN, et al. Investigating osteogenic differentiation in multiple myeloma using a novel 3D bone marrow niche model. Blood. 2014;124(22):3250–9.
12. Cao J, Wang Y, Dong R, Lin G, Zhang N, Wang J, et al. Hypoxia-induced WSB1 promotes the metastatic potential of osteosarcoma cells. Cancer Res. 2015;75(22):4839–51.
13. Deng J, Liu S, Li G, Zheng Y, Zhang W, Lin J, et al. pH-sensitive charge-conversion cinnamaldehyde polymeric prodrug micelles for effective targeted chemotherapy of osteosarcoma in vitro. Front Chem. 2023;11:1190596.
14. Wang J, Ferrena A, Zhang R, Singh S, Viscarret V, Al-Harden W, et al. Targeted inhibition of SCF(SKP2) confers anti-tumor activities resulting in a survival benefit in osteosarcoma. Oncogene. 2024;43(13):962–75.
15. Xu J, Shi Q, Lou J, Wang B, Wang W, Niu J, et al. Chordoma recruits and polarizes tumor-associated macrophages via secreting CCL5 to promote malignant progression. J Immunother Cancer. 2023;11(4):e006808.
16. Nie JH, Yang T, Li H, Li S, Li TT, Ye HS, et al. Frequently expressed glypican-3 as a promising novel therapeutic target for osteosarcomas. Cancer Sci. 2022;113(10):3618–32.
17. Visconti RJ, Kolaja K, Cottrell JA. A functional three-dimensional microphysiological human model of myeloma bone disease. J Bone Miner Res. 2021;36(10):1914–30.
18. Anloague A, Sabol HM, Kaur J, Khan S, Ashby C, Schinke C, et al. A novel CCL3-HMGB1 signaling axis regulating osteocyte RANKL expression in multiple myeloma. Haematologica. 2025;110(4):952–66.
19. Nelson EA, Walker SR, Xiang M, Weisberg E, Bar-Natan M, Barrett R, et al. The STAT5 inhibitor pimozide displays efficacy in models of acute myelogenous Leukemia driven by FLT3 mutations. Genes Cancer. 2012;3(7–8):503–11.
20. Subramaniam D, Angulo P, Ponnurangam S, Dandawate P, Ramamoorthy P, Srinivasan P, et al. Suppressing STAT5 signaling affects osteosarcoma growth and stemness. Cell Death Dis. 2020;11(2):149.

21. Zhang W, Qi L, Xu H, Yin C, Yu Z, Xu R, et al. Cooperative blockade of FLT3 and ALK synergistically suppresses growth of osteosarcoma. Oncogene. 2025;44(7):427–38.
22. Huang Y, Cao D, Zhang M, Yang Y, Niu G, Tang L, et al. Exploring the impact of PDGFD in osteosarcoma metastasis through single-cell sequencing analysis. Cell Oncol (Dordr). 2024;47(5):1715–33.
23. Walcott BP, Nahed BV, Mohyeldin A, Coumans JV, Kahle KT, Ferreira MJ. Chordoma: current concepts, management, and future directions. Lancet Oncol. 2012;13(2):e69–76.
24. Scognamiglio G, De Chiara A, Parafioriti A, Armiraglio E, Fazioli F, Gallo M, et al. Patient-derived organoids as a potential model to predict response to PD-1/PD-L1 checkpoint inhibitors. Br J Cancer. 2019;121(11):979–82.
25. Wu H, Li X, Zhang B, Liu P, Qi M, Du Y, et al. Single-cell sequencing reveals VEGFR as a potential target for CAR-T cell therapy in chordoma. Br J Cancer. 2024;130(10):1609–20.
26. Al Shihabi A, Davarifar A, Nguyen HTL, Tavanaie N, Nelson SD, Yanagawa J, et al. Personalized chordoma organoids for drug discovery studies. Sci Adv. 2022;8(7):eabl3674.
27. Jahanseir K, Folpe AL, Graham RP, Giannini C, Robinson SI, Sukov W, et al. Ewing sarcoma in older adults: a clinicopathologic study of 50 cases occurring in patients aged >/=40 years, with emphasis on histologic mimics. Int J Surg Pathol. 2020;28(4):352–60.
28. Maurer LM, Venier RM, Julian C, Bailey KM. Abstract A14: Exploiting DNA damage repair defects to enhance PD-L1 expression in Ewing sarcoma. Cancer Res. 2020;80(14_Supplement):A14.
29. Maurer LM, Daley JD, Mukherjee E, Venier RE, Julian CM, Bailey NG, et al. BRCA1-associated RING domain-1 (BARD1) loss and GBP1 expression enhance sensitivity to DNA damage in Ewing sarcoma. Cancer Res Commun. 2022;2(4):220–32.
30. Xu S, Tan S, Guo L. Patient-derived organoids as a promising tool for multimodal management of sarcomas. Cancers (Basel). 2023;15(17):4339.
31. Veys C, Benmoussa A, Contentin R, Duchemin A, Brotin E, Lafont JE, et al. Tumor suppressive role of miR-342-5p in human chondrosarcoma cells and 3D organoids. Int J Mol Sci. 2021;22(11):5590.
32. Suzuki R, Wakamatsu T, Yoshida K, Matsuoka Y, Takami H, Nakai S, et al. Genetic characterization of a novel organoid from human malignant giant-cell tumor. J Bone Oncol. 2023;41:100486.
33. Costa S, Rodrigues J, Vieira C, Dias S, Viegas J, Castro F, et al. Advancing osteosarcoma 3D modeling in vitro for novel tumor microenvironment-targeted therapies development. J Control Release. 2024;376:1068–85.
34. Rodrigues J, Sarmento B, Pereira CL. Osteosarcoma tumor microenvironment: the key for the successful development of biologically relevant 3D in vitro models. In Vitro Model. 2022;1(1):5–27.
35. Monteiro CF, Custodio CA, Mano JF. Bioengineering a humanized 3D tri-culture osteosarcoma model to assess tumor invasiveness and therapy response. Acta Biomater. 2021;134:204–14.
36. Brulin B, Nolan JC, Marangon T, Kovacevic M, Chatelais M, Meheust P, et al. Evaluation of the chemotherapy drug response using organotypic cultures of osteosarcoma tumours from mice models and canine patients. Cancers (Basel). 2021;13(19):4890.
37. He A, Huang Y, Cheng W, Zhang D, He W, Bai Y, et al. Organoid culture system for patient-derived lung metastatic osteosarcoma. Med Oncol. 2020;37(11):105.
38. Sarma MK, Saucedo A, Darwin CH, Felker ER, Umachandran K, Kohanghadosh D, et al. Noninvasive assessment of abdominal adipose tissues and quantification of hepatic and pancreatic fat fractions in type 2 diabetes mellitus. Magn Reson Imaging. 2020;72:95–102.
39. Gai T, Zhang H, Hu Y, Li R, Wang J, Chen X, et al. Sequential construction of vascularized and mineralized bone organoids using engineered ECM-DNA-CPO-based bionic matrix for efficient bone regeneration. Bioact Mater. 2025;49:362–77.

Chapter 6
Organoid Technology for Brain Disorders

Deniz Altunsu Kurt and Paulo Lizano

Abstract Organoids are three-dimensional (3D) cell cultures that replicate key structural and functional aspects of human organs. Derived from stem cells or tissue explants, organoids offer significant advantages over traditional two-dimensional (2D) cultures and animal models by more accurately simulating human physiology. Organoids have been successfully developed for various organs, including the brain, kidney, liver, and breast, providing valuable insights into organ biology, disease modeling, and drug discovery. Brain organoids, in particular, closely resemble the cellular composition and anatomical structure of the human brain, making them valuable models for studying brain development and neurological disorders. In this chapter, we focus on the application of brain organoids in disease modeling and discuss the current limitations of organoid technology.

Keywords Brain organoids · Disease modeling · Personalized medicine · Translational research

6.1 Introduction

Organoids are three-dimensional (3D) cell cultures, which allow the generation of organ-like structures from stem cells in Matrigel or basement membrane, combining with specific cytokines [45]. Organoids have specialized cell types arranged into the organ, so they can show specific organ function and physiology in vitro conditions. Since these organoids closely replicate human brain physiology, they offer a substantial advance over traditional methods such as two-dimensional cell cultures

D. A. Kurt (✉) · P. Lizano
Department of Psychiatry, University of Rochester Medical Center, Rochester, NY, USA

Department of Psychiatry, Harvard Medical School, Boston, MA, USA
e-mail: daltunsukurt@mgh.harvard.edu; paulo_lizano@urmc.rochester.edu

O. S. Somuncu (ed.), *Organoids for Preclinical Disease Modeling and Personalized Medicine*, Stem Cell Biology and Regenerative Medicine 78,
https://doi.org/10.1007/978-3-032-10024-5_6

and animal models. This innovation is particularly beneficial for studying neurological and neurodevelopmental disorders, as organoid models present stable genetic and phenotypic characteristics [6, 14, 17].

The brain is the most complex organ due to the intricate neuronal activity. It is involved in feeling, vision, memory, and other cognitive processes. Therefore, researchers face some difficulties in studying the human brain such as unique functional division, ethical concerns, and limited source of human brain tissue. So, animal models and 2D cell culture have been commonly used to understand the human brain physiology and pathology of diseases. However, it is challenging to demonstrate the developmental features of the human brain in animal models due to the distinct structure and function of human brain development, particularly in gene variation-related neurodegenerative diseases, including schizophrenia, autism spectrum disorder (ASD), Parkinson's disease (PD), and Alzheimer's disease (AD) [20, 54]. In contrast, stem cell technology allows a more reliable alternative to researchers by using human-induced pluripotent stem cells (hiPSC) to differentiate them into brain-like tissues, region-specific brain organoids, and organs. These three-dimensional structures capture essential features of neural architecture, resulting in a representation that is more closely aligned with human biology compared to standard animal models. The use of brain organoids contributes to a wide range of research applications such as drug screening and understanding human brain physiology, pathophysiology, and pharmacological features [9, 49]. In other words, brain organoids are a more preferable choice for personalized medicine. Organoids derived from patients allow researchers to assess specific drug responses, paving the way for customized treatment strategies. For example, organoids derived from medulloblastoma patients' tumors have been used to test cancer drugs [4]. Additionally, brain organoids have been grafted with glioblastoma to evaluate novel treatments and facilitate personalized medicine [46, 48].

Although brain organoids are promising models for studying human brain development and diseases, they possess some significant limitations structurally, functionally, and biologically. Structurally, brain organoids lack vascularization, which restricts nutrient and oxygen supply, leading to necrotic cores and limiting their size [1, 66]. Furthermore, these models often lack the intricate structural organization characteristic of the human brain, leading to considerable variability in their morphology, dimensions, and cellular composition [12]. Functionally, while organoids exhibit some neuronal activity, their neurons remain immature, and they struggle to form long-range connections necessary for higher cognitive functions [29]. The absence of sensory input and motor output further limits their ability to model real brain behavior [69]. From a biological perspective, the increasing sophistication of organoid models raises significant ethical concerns, particularly in relation to their potential to develop features associated with sentience. Furthermore, genetic and epigenetic variability affects reproducibility, and the lack of interactions with other systems, such as the immune and circulatory systems, prevents a comprehensive understanding of brain-body dynamics. Despite these challenges, advancements in vascularization, bioengineering, and computational modeling hold promise for overcoming these limitations, paving the way for more sophisticated and reliable brain organoid models in the future. Recent advancements in various technologies are being explored, such as advanced bioengineering, CRISPR-Cas9-based gene editing, computational modeling, and high-throughput screening, to enhance the authenticity and efficacy of brain organoid models.

In conclusion, brain organoids represent a transformative approach in the study of brain diseases and the acceleration of novel therapeutic strategies. While challenges persist in their application, the continued development of organoid technology holds considerable promise for advancing neurotherapy and improving drug discovery for a variety of brain-related disorders. As this field evolves, it may ultimately lead to significant breakthroughs in personalized medicine and our understanding of complex neurological conditions.

6.2 Applications of Brain Organoids as Disease Models

Animal models or postmortem brain tissues have been commonly used for understanding the human brain physiology and pathophysiology of neurological disorders. Although postmortem studies contribute valuable insights into brain pathology, they are inherently limited to deceased individuals, making it difficult to observe dynamic disease progression and developmental processes [32, 62]. On the other hand, animal models offer genetic and cellular insights into disease mechanisms but fail to fully recapitulate human brain complexity due to species-specific differences. Moreover, ethical concerns surrounding animal research further necessitate alternative models. To minimize these limitations, hiPSCs have emerged as a transformative tool, enabling the generation of brain organoids that model early brain development and neurological disorders, including autism spectrum disorder, Alzheimer's disease, and Zika virus-induced microcephaly (Fig. 6.1) [10, 31, 41, 63]. These patient-derived organoids provide for precision medicine approaches, offering a platform for personalized therapy and biobanking for disease modeling. Additionally, advanced technologies such as CRISPR/Cas9 gene editing have enhanced the ability to create targeted disease models, furthering the potential of brain organoids in neurological research [42].

Brain organoids have been used as a transformative tool in neuroscience, providing an innovative platform to investigate the neurobiology of brain disorders such as ASD, AD, PD, and schizophrenia, by replicating key aspects of human neural structure and function. The ability of brain organoids to closely mimic human neural architecture has made them invaluable tools for studying neurodevelopmental and degenerative diseases.

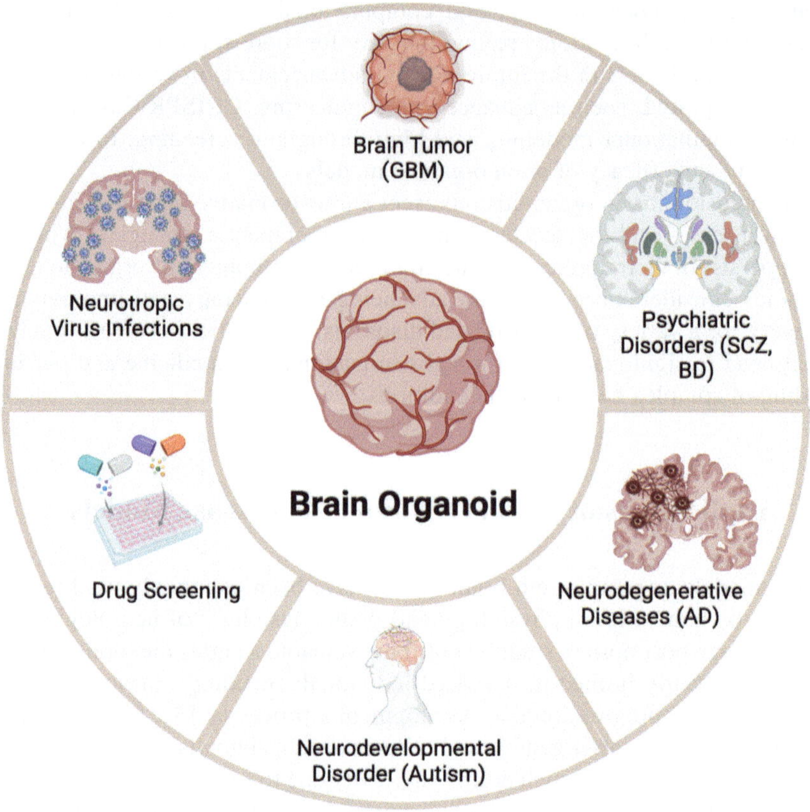

Fig. 6.1 Applications of brain organoids in biomedical research. (Figure is generated by BioRender (https://BioRender.com/fz3tsnj))

6.2.1 Autism Spectrum Disorders

Autism spectrum disorder (ASD) is a neurodevelopmental disorder, which is caused by different risk factors, including genetic mutation, epigenetic alterations, and environmental factors [2]. Brain organoids allow scientists to examine early neural development, genetic mutations, and cellular interactions in a controlled environment. By mimicking aspects of the developing human brain, organoids provide insights into the molecular and cellular mechanisms underlying ASD. Recent studies have utilized organoids to investigate genetic mutations associated with autism, analyze synaptic and neuronal defects, and explore potential therapeutic interventions. For example, research on CHD2 haploinsufficiency has shown to be important for cortical organoid maturation, linking autolysosomal pathways to neurodevelopmental defects [36]. Similarly, mutations in Neuroligin-3 (NLGN3 R451C) have been found to promote excessive astroglia proliferation in organoids, potentially affecting neuronal support and signaling [16]. Another study on autism patient-derived organoids indicates that an immature choroid plexus may contribute to abnormal cerebrospinal fluid composition and neurodevelopmental alterations [59]. Transcriptomic analyses of macrocephalic autism organoids suggest the involvement of Ndel1 and cytoskeletal proteins in neural structural abnormalities [57]. Additionally, the deficiency of FABP7 in autism organoids has been shown to trigger premature neural differentiation, leading to behavioral deficits [25]. The knockdown of the GBX2 gene in brain organoids has also been found to disrupt thalamic differentiation, which could be relevant to sensory processing deficits in ASD [35]. Other studies have revealed that AUTS2 disruption in organoids leads to hyperactivation of the WNT/β-catenin pathway, causing defects in neuronal differentiation and growth [24], while epigenetic dysregulation in ASD organoids has been linked to altered neural development and behavioral phenotypes [23, 27].

Interestingly, organoid-based studies have also demonstrated sex-specific differences in neural connectivity in ASD, suggesting a possible influence of hormonal or genetic factors [47]. Moreover, neuronal lineage tracing in cortical organoids has identified distinct progenitor subtypes contributing to neuronal diversity and pathology in ASD [7, 67]. Collectively, these findings emphasize the immense potential of organoid models in unraveling the complex biology of autism, providing valuable insights into cellular abnormalities, neurotransmitter imbalances, and novel therapeutic targets. As organoid technology continues to advance, future research may further enhance our understanding of ASD by refining these models to better replicate neural circuit formation and brain connectivity.

6.2.2 Alzheimer's Disease

Alzheimer's disease is the most prevalent neurodegenerative disease, which is identified by progressive reduction, cognitive decline, and behavioral impairment. Brain organoids have emerged as an essential tool in AD research, providing a three-dimensional model that closely mimics the complex cellular interactions of the human brain. These models enable researchers to investigate critical aspects of AD pathology, including amyloid-beta (Aβ) aggregation, tau hyperphosphorylation, and neuroinflammation. Studies have demonstrated that brain organoids derived from AD patient cells exhibit increased Aβ deposits and tau tangles, similar to those observed in postmortem brains of AD patients [60]. Additionally, researchers have successfully incorporated microglia into brain organoids, which has allowed them to explore the role of neuroinflammation in AD progression [33, 51].

A significant breakthrough was the development of organoids with a functional blood-brain barrier, aiding the study of vascular contributions to AD [5]. Moreover, transcriptomic and epigenetic studies using brain organoids have provided insights into disease-associated gene expression changes, particularly in pathways related to synaptic function and neuronal survival [44, 64]. Organoid-based models have also been instrumental in testing potential therapeutic approaches, such as kinase inhibitors targeting tau aggregation and drug screening for compounds that enhance neuronal resilience [18]. Another study highlighted the impact of atmospheric heavy metals on AD pathology, demonstrating that exposure to environmental toxins could exacerbate neurodegeneration in organoid models [55]. Furthermore, a novel approach using iPSC-derived microglial cells has shown their protective role against neuronal damage in long-term cultured brain organoids, suggesting potential therapeutic avenues [11]. These advancements underscore the critical role of brain organoids in advancing our understanding of AD pathogenesis, offering a promising platform for studying disease mechanisms and accelerating drug discovery.

6.2.3 *Parkinson's Disease*

Brain organoids have become a revolutionary tool in Parkinson's disease (PD) research, providing a three-dimensional model that better mimics the human brain compared to traditional 2D cultures and animal models. PD, a neurodegenerative disorder characterized by the progressive loss of dopaminergic neurons in the substantia nigra, has been challenging to study due to the complexity of human brain function. Recent advancements have allowed scientists to develop midbrain-specific organoids that contain dopaminergic neurons, enabling the study of key disease mechanisms such as mitochondrial dysfunction, neuroinflammation, and synaptic degeneration. For example, mutations in the Miro1 gene, a key regulator of mitochondrial transport, contribute to dopaminergic neuron loss in midbrain organoids, highlighting mitochondrial dysfunction as a major factor in PD progression [8]. Furthermore, studies have integrated microglia (brain immune cells) into organoids to investigate the role of neuroinflammation in neuronal damage using iPSC-derived microglia to study inflammatory responses in PD [28, 39].

Another breakthrough is the use of organoids for personalized medicine, where patient-derived organoids allow researchers to test different drugs and assess their effectiveness in specific genetic backgrounds. This approach is particularly useful for screening potential neuroprotective compounds and evaluating gene therapies using CRISPR-based editing. Additionally, regenerative medicine is exploring how dopaminergic neuron transplantation from organoids could serve as a future treatment strategy for PD by replenishing lost neurons [22, 26]. As research advances, efforts are underway to improve organoid models through the incorporation of vascular networks, integration with organ-on-a-chip platforms, and exploration of their potential in biocomputing applications—particularly for investigating neural networks implicated in Parkinson's disease [13, 71]. With ongoing advancements, brain organoids are set to revolutionize Parkinson's disease research by providing more accurate, patient-specific models that can accelerate drug discovery, improve regenerative therapies, and ultimately bring us closer to effective treatments for this devastating disorder.

6.2.4 Schizophrenia

Brain organoids have become a powerful tool in neuropsychiatric disorders, offering a three-dimensional model to study the molecular and cellular abnormalities underlying the pathology. Studies using patient-derived brain organoids have identified key mitochondrial autophagy genes implicated in schizophrenia, providing insights into the role of disrupted mitophagy in disease pathology [34]. Additionally, specific cytoskeletal proteins and neuronal markers, such as MAP2, TUBB3, and GAP43, were found to be depleted in brain organoids derived from schizophrenia patients in proteomic results [43]. Multiomic approaches integrating single-cell RNA sequencing have further demonstrated unique gene expression patterns in schizophrenia organoids, particularly in pathways associated with neurodevelopment and synaptic function [37]. Researchers have also used forebrain organoids to investigate the effects of Neurexin-1 (NRXN1) deletions, a genetic variant linked to schizophrenia, revealing its impact on early brain development and connectivity [40, 53]. Another study analyzed single-cell RNA sequencing data from schizophrenia organoids, identifying disruptions in mitophagy and energy metabolism that could contribute to neuronal dysfunction [68]. These findings underline the critical role of brain organoids in advancing our understanding of schizophrenia, enabling researchers to model disease-relevant genetic and molecular changes, explore neurodevelopmental disruptions, and test potential therapeutic interventions.

6.2.5 Brain Tumors

Brain organoids have emerged as powerful tools in brain tumor research, offering three-dimensional models that closely mimic human brain tissue. These organoids provide critical insights into tumor development, invasion, and response to therapies. Recent studies highlight their role in glioblastoma research, where extracellular vesicles from tumor cells facilitate invasion into organoid structures, mimicking real tumor behavior [52, 61, 70]. Organoids also help in understanding the heterogeneity of the tumor microenvironment, enhancing personalized medicine approaches. Additionally, patient-derived tumor spheroids and iPSC-derived cerebral organoids are being utilized to study metastasis and tumor-stroma interactions, aiding in drug discovery [3, 30]. Advanced models, such as organoid-on-a-chip systems, further refine our understanding of tumor progression and drug responses. Studies have also demonstrated their potential in tracking glioblastoma cell-state heterogeneity and in building biobanks for glioma research [38, 65]. Overall, brain organoids are revolutionizing neuro-oncology by providing an accurate, patient-specific platform for studying brain tumors, testing novel treatments, and bridging the gap between in vitro experiments and clinical applications.

6.2.6 Neurotropic Virus Infections

Neurotropic viruses are pathogens that specifically target the nervous system, often causing severe neurological damage. The brain plays a critical role in both the pathogenesis and immune response to these infections. These 3D structures mimic human brain development and allow for the investigation of viral tropism, pathogenesis, and cellular responses [19, 58]. Studies have utilized brain organoids to examine infections caused by various viruses, including Zika, SARS-CoV-2, HIV, HSV-1, and influenza [15, 21, 56]. Research has revealed that different viruses can affect organoid size, induce apoptosis, and trigger inflammatory responses [56]. Brain organoids offer advantages over traditional animal models and 2D cultures, providing insights into human-specific neurodevelopmental processes and virus-host interactions [50]. While limitations exist, such as heterogeneity in protocols and lack of certain cell types, brain organoids represent a promising platform for understanding neurotropic viral infections and screening potential therapeutics [19, 58].

6.3 Future Directions of Brain Organoids

The future of brain organoids is set to revolutionize neuroscience, regenerative medicine, and disease modeling by addressing existing challenges and expanding their applications. Scientists are working on enhancing the physiological accuracy of these models by integrating vascularization, immune system components, and functional neural circuits to better mimic the human brain. One of the primary goals is to improve organoid maturation and longevity, allowing for more advanced studies on neurodevelopment, neurodegenerative diseases, and therapeutic interventions [12].

A significant limitation of current brain organoids is the lack of a vascular system, which limits their growth and prevents the formation of larger, more complex structures. Future developments aim to incorporate endothelial cells to create self-sustaining vascular networks, which will enable better nutrient and oxygen exchange [69]. Additionally, integrating microglia and other immune components into brain organoids will allow scientists to study neuroimmune interactions, which play a crucial role in neuroinflammatory and neurodegenerative disorders.

Another major advancement will be the development of more sophisticated brain organoids that mimic functional neural circuits, allowing for the study of synaptic connectivity, neural oscillations, and even cognitive-like functions. New approaches, such as fusing multiple brain region-specific organoids (e.g., cortical, hippocampal, and thalamic organoids), will help model brain-wide communication more accurately [12]. This will enhance research into psychiatric disorders, neurodevelopmental diseases, and learning-related mechanisms. The combination of brain organoids with microfluidic "organ-on-a-chip" platforms will significantly improve their ability to model dynamic physiological conditions. These systems will enable precise control over nutrient delivery, waste removal, and mechanical stimulation, closely simulating real brain environments. This will be particularly useful in studying drug responses and brain disorders in a more physiologically relevant context. A groundbreaking field emerging from brain organoid research is Organoid Intelligence (OI), the use of living neural tissue to perform computational tasks. Scientists are exploring the potential of brain organoids in biocomputing, where these structures could be used to create biological processing units capable of learning and problem-solving. While still in early stages, this technology could redefine how we approach artificial intelligence by merging biological and synthetic computing. With the advancement of patient-specific brain organoids, researchers can develop personalized models to study neurological diseases such as Alzheimer's, Parkinson's, and epilepsy. These organoids, derived from induced pluripotent stem cells (iPSCs), will allow clinicians to test drug responses on patient-specific tissue before administering treatments, leading to precision medicine approaches. Additionally, brain organoids will be used for large-scale drug screening, helping identify new therapeutic compounds more efficiently. Gene-editing technologies like CRISPR-Cas9 are being used to introduce specific genetic mutations into brain organoids, allowing scientists to model neurodevelopmental and neurodegenerative disorders with unprecedented accuracy. This approach will provide valuable insights into conditions such as autism, schizophrenia, and rare genetic brain diseases, leading to targeted gene therapies.

As brain organoids become more complex and develop cognitive-like functions, ethical concerns about their moral status and potential consciousness are gaining attention. Researchers are working closely with ethicists and policymakers to establish guidelines on the permissible level of brain organoid development and their applications in research [12]. This discussion is particularly relevant as organoids begin to exhibit neural activity patterns similar to human brain waves.

6.4 Conclusion

The future of brain organoids is full of exciting possibilities, ranging from medical applications in neurodegenerative diseases to groundbreaking innovations in biocomputing and artificial intelligence. By enhancing vascularization, integrating immune components, and improving neural connectivity, researchers will create more physiologically relevant brain models. The fusion of brain organoids with cutting-edge technologies like microfluidics, CRISPR, and artificial intelligence will push the boundaries of neuroscience and medicine. However, as these models become more advanced, ethical and regulatory considerations will need to be addressed to ensure responsible use. With continuous advancements, brain organoids are set to redefine our understanding of the human brain and revolutionize neurological research and treatment strategies.

References

1. Alameri A. Microfluidic-based organoid vascularization in PDMS devices replicated from 3D printed molds. McGill University (Canada). 2022. https://search.proquest.com/openview/1f02fc2583238e92a711ddf02fc66894/1?pq-origsite=gscholar&cbl=18750&diss=y.
2. Almandil NB, Alkuroud DN, AbdulAzeez S, AlSulaiman A, Elaissari A, Borgio JF. Environmental and genetic factors in autism spectrum disorders: special emphasis on data from Arabian studies. Int J Environ Res Public Health. 2019;16:658.
3. Anderle N. Enhancing precision oncology: preclinical assessment of individual drug treatment susceptibility using patient-derived 3D microtumor and immune cell co-cultures. PhD thesis, Universität Tübingen. 2024. https://tobias-lib.ub.uni-tuebingen.de/xmlui/handle/10900/153006.
4. Ballabio C, Anderle M, Gianesello M, Lago C, Miele E, Cardano M, Aiello G, Piazza S, Caron D, Gianno F. Modeling medulloblastoma in vivo and with human cerebellar organoids. Nat Commun. 2020;11:583.
5. Birol M. BPS2025-early deviations in cell-to-cell communication in neurodegeneration: systems-level biophysics using brain organoid models. Biophys J. 2025;124:378a–9a.
6. Bock M, Hong SJ, Zhang S, Yu Y, Lee S, Shin H, Choi BH, Han I. Morphogenetic designs, and disease models in central nervous system organoids. Int J Mol Sci. 2024;25:7750.
7. Bury LA, Fu S, Wynshaw-Boris A. Neuronal lineage tracing from progenitors in human cortical organoids reveals mechanisms of neuronal production, diversity, and disease. Cell Rep. 2024;43:114862. https://www.cell.com/cell-reports/fulltext/S2211-1247(24)01213-0.
8. Chemla A, Arena G, Sacripanti G, Barmpa K, Zagare A, Garcia P, Gorgogietas V, Antony P, Ohnmacht J, Baron A. Parkinson's disease mutant Miro1 causes mitochondrial dysfunction and dopaminergic neuron loss. Brain. 2025;148:3607–22.
9. Chen HI, Song H, Ming G. Applications of human brain organoids to clinical problems. Dev Dyn. 2019;248:53–64. https://doi.org/10.1002/dvdy.24662.
10. Chen H, Jin X, Li T, Ye Z. Brain organoids: establishment and application. Front Cell Dev Biol. 2022;10:1029873.
11. Chen X, Sun G, Feng L, Tian E, Shi Y. Human iPSC-derived microglial cells protect neurons from neurodegeneration in long-term cultured adhesion brain organoids. Commun Biol. 2025;8:30.
12. Cheng K, Kshirsagar A, Nixon J, Lau J, Yang K, Sawa A, Kathuria A. Model systems for emulating human tissue and physiology in psychiatric research. Front Neurosci. 2025;19:1527826.
13. Cheruku GR, Wilson CV, Raviendran S, Xiao Q. Recent advances and future perspectives in vascular organoids and vessel-on-cChip. Organ. 2024;3:203–46.
14. Coronel R, García-Moreno E, Siendones E, Barrero MJ, Martínez-Delgado B, Santos-Ocaña C, Liste I, Cascajo-Almenara MV. Brain organoid as a model to study the role of mitochondria in neurodevelopmental disorders: achievements and weaknesses. Front Cell Neurosci. 2024;18:1403734.
15. Crawford G, Soper O, Kang E, Berg DA. Advancing insights into virus-induced neurodevelopmental disorders through human brain organoid modelling. Expert Rev Mol Med. 2024;27:e1.
16. Dang R, Dalmia M, Ma Z, Jin M, Aluru K, Mirabella VR, Papetti AV, Cai L, Jiang P. Neuroligin-3 R451C induces gain-of-function gene expression in astroglia in an astroglia-enriched brain organoid model. Cell Regen. 2025;14:1. https://doi.org/10.1186/s13619-024-00219-5.
17. de Lagrán MM, Bascón-Cardozo K, Dierssen M. Neurodevelopmental disorders: 2024 update. Free Neuropathol. 2024;5:5–20.
18. Debbi E. Impact of atmospheric heavy metals on synaptic development via hiPSC-based 2D and 3D brain models. 2025. https://tesidottorato.depositolegale.it/handle/20.500.14242/189703.
19. Depla JA, Mulder LA, de Sá RV, Wartel M, Sridhar A, Evers MM, Wolthers KC, Pajkrt D. Human brain organoids as models for central nervous system viral infection. Viruses. 2022;14:634.

20. Dey A, Ghosh S, Majhi S, Hoque AN, Chakrabarti A. Systems biology approaches as a tool in understanding neurodevelopmental disorders: some case studies. In: Joshi S, Ray RR, Nag M, Lahiri D, editors. Systems biology approaches: prevention, diagnosis, and understanding mechanisms of complex diseases. Singapore: Springer Nature; 2024. p. 511–36. https://doi.org/10.1007/978-981-99-9462-5_20.
21. Fan W, Christian KM, Song H, Ming G. Applications of brain organoids for infectious diseases. J Mol Biol. 2022;434:167243.
22. Fiorenzano A, Sozzi E, Parmar M, Storm P. Dopamine neuron diversity: recent advances and current challenges in human stem cell models and single cell sequencing. Cells. 2021;10:1366.
23. Forsberg SL, Ilieva M, Maria Michel T. Epigenetics and cerebral organoids: promising directions in autism spectrum disorders. Transl Psychiatry. 2018;8:14.
24. Geng Z, Tai YT, Wang Q, Gao Z. AUTS2 disruption causes neuronal differentiation defects in human cerebral organoids through hyperactivation of the WNT/β-catenin pathway. Sci Rep. 2024;14:19522.
25. Han X, He Y, Wang Y, Hu W, Chu C, Huang L, Hong Y, Han L, Zhang X, Gao Y, Lin Y, Ma H, Shen H, Ke X, Liu Y, Hu Z. Deficiency of FABP7 triggers premature neural differentiation in idiopathic normocephalic autism organoids. Adv Sci. 2025;12:2406849. https://doi.org/10.1002/advs.202406849.
26. Harary PM, Jgamadze D, Kim J, Wolf JA, Song H, Ming G, Cullen DK, Chen HI. Cell replacement therapy for brain repair: recent progress and remaining challenges for treating Parkinson's disease and cortical injury. Brain Sci. 2023;13:1654.
27. Herrera ML, Paraíso-Luna J, Bustos-Martínez I, Barco Á. Targeting epigenetic dysregulation in autism spectrum disorders. Trends Mol Med. 2024;30:1028–46. https://www.cell.com/trends/molecular-medicine/abstract/S1471-4914(24)00162-X.
28. Hong Y, Dong X, Chang L, Xie C, Chang M, Aguilar JS, Lin J, Lin J, Li QQ. Microglia-containing cerebral organoids derived from induced pluripotent stem cells for the study of neurological diseases. Iscience. 2023;26:106267. https://www.cell.com/iscience/fulltext/S2589-0042(23)00344-9.
29. Kanupriya K, Verma SP, Sharma V, Mishra I, Mishra R. Advances in human brain organoids: methodological innovations and future directions for drug discovery. Curr Drug Res Rev. 2025. Retrieved March 1 2025, from https://pubmed.ncbi.nlm.nih.gov/39976096/.
30. Karakaya M, Obakan Yerlikaya P. Recent in vitro models and tissue engineering strategies to study glioblastoma. Biotech Stud. 2024;33:52–66. https://doi.org/10.38042/biotechstudies.1463814.
31. Kataoka M, Ishida S, Kobayashi C, Lee T-L, Sawai T. Evaluating neuroprivacy concerns in human brain organoid research. Trends Biotechnol. 2024;43:491–3. https://www.cell.com/trends/biotechnology/fulltext/S0167-7799(24)00245-2.
32. Krassner MM, Kauffman J, Sowa A, Cialowicz K, Walsh S, Farrell K, Crary JF, McKenzie AT. Postmortem changes in brain cell structure: a review. Free Neuropathol. 2023;4:4–10.
33. Kuhn MK, Kang RY, Kim C, Tagay Y, Morris N, Tabdanov ED, Elcheva IA, Proctor EA. Dynamic neuroinflammatory profiles predict Alzheimer's disease pathology in microglia-containing cerebral organoids. bioRxiv. 2024;2023:2023.11.16.567220.
34. Kun L, Yongmei LI, Chenglong SHI, Yilan C, Zhang L, Wei Y, Xiufeng XU. Identify key mitochondrial autophagy genes in schizophrenia through integrated bioinformatics approaches. J Kunming Medical University. 2025;46:23–35.
35. Lee Y, Choe MS. Determining the effect of autism spectrum disorder associated gene GBX2 deficiency on thalamic differentiation in human organoid models. J High School Sci. 2024;8:65–85.
36. Leonardi O, Lewerissa E, Aprile D, Barakat TS, Culotta L, Deng R, Haazen L, Linda K, Majoie M, Mordelt A, et al. CHD2 dosage ties autolysosomal pathway to cortical maturation in disease and evolution. bioRxiv. 2025;2025–01. https://doi.org/10.1101/2025.01.21.634145.

37. Malwade S. Multiomic modelling of schizophrenia using patient-derived brain organoids. PhD thesis, Karolinska Institutet. 2024. https://openarchive.ki.se/articles/thesis/Multiomic_modelling_of_schizophrenia_using_patient-derived_brain_organoids/27269790.
38. Mangena V, Chanoch-Myers R, Sartore R, Paulsen B, Gritsch S, Weisman H, Hara T, Breakefield XO, Breyne K, Regev A. Glioblastoma cortical organoids recapitulate cell-state heterogeneity and intercellular transfer. Cancer Discov. 2025;15:299–315.
39. Mrza MA, He J, Wang Y. Integration of iPSC-derived microglia into brain organoids for neurological research. Int J Mol Sci. 2024;25:3148.
40. Murphy A. Neurodevelopmental impact of patient-specific NRXN1 deletions in forebrain organoids. Master's thesis, Icahn School of Medicine at Mount Sinai. 2021. https://search.proquest.com/openview/0b51ecd15acbc3e447d330814d40665a/1?pq-origsite=gscholar&cbl=18750&diss=y.
41. National Academies of Sciences & Medicine. The emerging field of human neural organoids, transplants, and chimeras: science, ethics, and governance. 2021. https://nap.nationalacademies.org/catalog/26078/the-emerging-field-of-human-neural-organoids-transplants-and-chimeras.
42. Nojadeh JN, Eryilmaz NSB, Ergüder Bİ. CRISPR/Cas9 genome editing for neurodegenerative diseases. EXCLI J. 2023;22:567.
43. Notaras M, Lodhi A, Fang H, Greening D, Colak D. The proteomic architecture of schizophrenia iPSC-derived cerebral organoids reveals alterations in GWAS and neuronal development factors. Transl Psychiatry. 2021;11:541.
44. Nott A, Holtman IR. Genetic insights into immune mechanisms of Alzheimer's and Parkinson's disease. Front Immunol. 2023;14:1168539.
45. Pagliaro A, Artegiani B, Hendriks D. Emerging approaches to enhance human brain organoid physiology. Trends Cell Biol. 2025;35:483–99. https://www.cell.com/trends/cell-biology/fulltext/S0962-8924(24)00254-X
46. Pamies D, Zurich M-G, Hartung T. Organotypic models to study human glioblastoma: studying the beast in its ecosystem. Iscience. 2020;23:101633. https://www.cell.com/iscience/fulltext/S2589-0042(20)30825-7
47. Pavlinek A, Adhya D, Tsompanidis A, Warrier V, Vernon AC, Lancaster M, Mill J, Srivastava DP, Baron-Cohen S. Using organoids to model sex differences in the human brain. Biol Psychiatry Glob Open Sci. 2024;4:100343.
48. Plummer S, Wallace S, Ball G, Lloyd R, Schiapparelli P, Quiñones-Hinojosa A, Hartung T, Pamies D. A human iPSC-derived 3D platform using primary brain cancer cells to study drug development and personalized medicine. Sci Rep. 2019;9:1407.
49. Qian X, Song H, Ming G. Brain organoids: advances, applications and challenges. Development. 2019;146:dev166074.
50. Ramani A, Pranty A-I, Gopalakrishnan J. Neurotropic effects of SARS-CoV-2 modeled by the human brain organoids. Stem Cell Reports. 2021;16:373–84.
51. Rubio AD, Hamilton L, Bausch M, Jin M, Papetti A, Jiang P, Yelamanchili SV. A comprehensive review on utilizing human brain organoids to study neuroinflammation in neurological disorders. J NeuroImmune Pharmacol. 2025;20:23. https://doi.org/10.1007/s11481-025-10181-x.
52. Schiera G, Di Liegro CM, Vento F, Di Liegro I. Role of extracellular vesicles in the progression of brain tumors. Biology. 2024;13:586.
53. Sebastian R, Jin K, Pavon N, Bansal R, Potter A, Song Y, Babu J, Gabriel R, Sun Y, Aronow B. Schizophrenia-associated NRXN1 deletions induce developmental-timing-and cell-type-specific vulnerabilities in human brain organoids. Nat Commun. 2023;14:3770.
54. Sevimoglu T, Bal T, Özdemir Ç. Brain diseases and disorders. In: Bioinformatics of the brain. London: CRC Press; 2024. p. 1–41. https://www.taylorfrancis.com/chapters/edit/10.1201/9781003461906-1/brain-diseases-disorders-tuba-sevimoglu-tugba-bal-%C3%A7a%C4%9Fla-%C3%B6zdemir.
55. Shin YJ. Modeling perfusable human brain microvasculature: multimodal investigation of cerebrovasculature in dementia, trauma and development. PhD thesis. 2025. https://digital.lib.washington.edu/researchworks/items/58ab15cf-c989-49e4-9b52-0d583c74b655.

56. Su X, Yue P, Kong J, Xu X, Zhang Y, Cao W, Fan Y, Liu M, Chen J, Liu A. Human brain organoids as an in vitro model system of viral infectious diseases. Front Immunol. 2022;12:792316.
57. Susaimanickam PJ, Kiral FR, Park I-H. Region specific brain organoids to study neurodevelopmental disorders. Int J Stem Cells. 2022;15:26–40.
58. Swingler M, Donadoni M, Bellizzi A, Cakir S, Sariyer IK. iPSC-derived three-dimensional brain organoid models and neurotropic viral infections. J Neurovirol. 2023;29:121–34. https://doi.org/10.1007/s13365-023-01133-3.
59. Tanabe M, Saito Y, Takasaki A, Nakano K, Suzuki C, Kawamura N, Hattori A, Nagashima S, Oikawa M, Yanagi S. Role of immature choroid plexus in the pathology of autism spectrum disorder. Cell Rep. 2025;44:115133.
60. Tian C, Ao Z, Cerneckis J, Cai H, Chen L, Niu H, Takayama K, Kim J, Shi Y, Gu M. Understanding monocyte-driven neuroinflammation in Alzheimers disease using human brain organoid microphysiological systems. bioRxiv. 2025;2025:2025.02.16.638539.
61. Ural EE, Toomajian V, Hoque Apu E, Veletic M, Balasingham I, Ashammakhi N, Kanada M, Contag CH. Visualizing extracellular vesicles and their function in 3D tumor microenvironment models. Int J Mol Sci. 2021;22:4784.
62. Vornholt E, Luo D, Qiu W, McMichael GO, Liu Y, Gillespie N, Ma C, Vladimirov VI. Postmortem brain tissue as an underutilized resource to study the molecular pathology of neuropsychiatric disorders across different ethnic populations. Neurosci Biobehav Rev. 2019;102:195–207.
63. Wang H. Modeling neurological diseases with human brain organoids. Front Synaptic Neurosci. 2018;10:15.
64. Wang Y-W, Hu N, Li X-H. Genetic and epigenetic regulation of brain organoids. Front Cell Dev Biol. 2022;10:948818.
65. Wen Y, Bai H, Li Q, Huang S, Jia X, Pan G, Yao H. A biobank of patient derived glioma organoids. Sci China Life Sci. 2024;67:2780–2. https://doi.org/10.1007/s11427-024-2632-0.
66. Wörsdörfer P, Ergün S. The impact of oxygen availability and multilineage communication on organoid maturation. Antioxid Redox Signal. 2021;35:217–33. https://doi.org/10.1089/ars.2020.8195.
67. Yang G, Shcheglovitov A. Probing disrupted neurodevelopment in autism using human stem cell-derived neurons and organoids: an outlook into future diagnostics and drug development. Dev Dyn. 2020;249:6–33. https://doi.org/10.1002/dvdy.100.
68. Yang W, Lian K, Ye J, Cheng Y, Xu X. Analyses of single-cell and bulk RNA sequencing combined with machine learning reveal the expression patterns of disrupted mitophagy in schizophrenia. Front Psychol. 2024;15:1429437.
69. Yaron A, Zhang Z, Akita D, Shiramatsu TI, Chao Z, Takahashi H. Dissociated neuronal cultures as model systems for self-organized prediction. arXiv:2501.18772. 2025. https://doi.org/10.48550/arXiv.2501.18772.
70. Zhang Y, Lu A, Zhuang Z, Zhang S, Liu S, Chen H, Yang X, Wang Z. Can organoid model reveal a key role of extracellular vesicles in tumors? A comprehensive review of the literature. Int J Nanomedicine. 2023;18:5511–27. https://doi.org/10.2147/IJN.S424737.
71. Zhu J, Ji L, Chen Y, Li H, Huang M, Dai Z, Wang J, Xiang D, Fu G, Lei Z. Organoids and organs-on-chips: insights into predicting the efficacy of systemic treatment in colorectal cancer. Cell Death Dis. 2023;9:72.

Chapter 7
Investigating Pediatric Diseases with Organoid Systems: When Pediatric Oncology Meets Organoid Technology

Fulden Aycan

Abstract Pediatric cancers present unique biological challenges and remain a leading cause of disease-related death in children. While treatment advances have improved outcomes, progress is limited for many high-risk and relapsed tumors. Organoid technology—three-dimensional cultures derived from patient or stem cells—offers a powerful new platform to model pediatric tumors more accurately. This chapter explores the convergence of pediatric oncology and organoid technology, highlighting current applications, emerging models, and the potential to revolutionize both translational research and clinical care for children with cancer.

Keywords Pediatric cancers · Pediatric organoid technology · Pediatric oncology

7.1 Introduction

Pediatric cancers differ quite significantly from adult cancers. While childhood cancers are rare, affecting approximately 15 per 100,000 children every year, adult cancers occur more frequently [1]. Adult tumors are most commonly carcinomas derived from highly differentiated epithelial tissues, such as in breast, lung, colon, and prostate cancers. Pediatric cancers girdle a heterogeneous cluster of diseases that can be broadly subclassified into leukemias, lymphomas, brain and non-central nervous system tumors, sarcomas, and additional rare cancers [2]. The genomic landscape of childhood cancers also differs from adult cancers. While numerous mutational "hits" are frequently observed in adult cancers, pediatric cancer-driving point mutations are enriched in genes that encode epigenetic machinery [3]. Pediatric cancer is characterized by a low mutation burden and mutations are

F. Aycan (✉)
Stead Family Children's Hospital, University of Iowa Health Care Medical Center, Iowa City, IA, USA
e-mail: fulden-aycan@uiowa.edu

O. S. Somuncu (ed.), *Organoids for Preclinical Disease Modeling and Personalized Medicine*, Stem Cell Biology and Regenerative Medicine 78,
https://doi.org/10.1007/978-3-032-10024-5_7

concentrated in genes associated with the epigenetic machinery and developmental pathways, such as Notch, WNT (Wingless), SHH (Sonic Hedgehog), and TGF-beta. Another distinctive element of childhood cancer is the high prevalence of fusion oncoproteins, for example, the EWS-FLI1 gene in Ewing sarcoma (ES) and the PAX-FOXO1 in rhabdomyosarcoma (RMS) [4]. The genesis of chemotherapy and its application in childhood cancers is often cited as the modern success story of medical research. Cure rates for childhood cancers have increased from <25% in the pre-chemotherapy era to about 80% in recent decades [5, 6]. It is essential to acknowledge that pediatric cancer is still the leading cause of death by disease for children in western countries, with 20% of children not responding or relapsing after first-line treatment. The lack of efficient second-line therapies also contributes to poor outcomes [2]. Most of the children who survive cancer suffer from long-term sequelae including organ toxicity, mental disabilities, and secondary cancers [7].

7.2 Pediatric Cancers and Organoids

Organoids represent a promising approach for creating patient-derived in vitro cancer models that closely recapitulate the overall pathophysiological features of natural tumorigenesis, including intra-tumoral heterogeneity and plasticity. Organoids can be described as self-organized 3D tissues, capable of mimicking an organ's key functional, structural, and biological complexity. Organoids could provide an unprecedented tool to unravel the biology of pediatric cancers and improve their therapeutic management [8, 9]. Organoid technology facilitates the generation of extensive living material collections for research and overcomes the challenges posed by the relative rarity and limited sample sizes of these tumors (Fig. 7.1). Although this technology has been frequently applied to adult cancers, it has rarely been used for childhood cancers, due to the notable delay in transfer. Compared to the dozen in adult oncology, only a few protocols have been documented for tumor organoids obtained from pediatric patients [10].

7.3 Pediatric Brain Tumors

Brain tumors occur more frequently in adults when compared with children. The most common solid tumors in children are brain tumors, which only cause 25% of pediatric cancers. Approximately 60% of pediatric brain tumors arise in the infratentorial region, predominantly in the posterior fossa, encompassing the brain stem and cerebellum [11, 12]. Despite notable progress in establishing organoid models for adult brain tumors like glioblastoma (GBM) or low-grade gliomas (LGGs), there is a significant gap in equivalent research for pediatric brain tumors. Currently, the use of organoid models for pediatric brain tumors has been circumscribed, with the

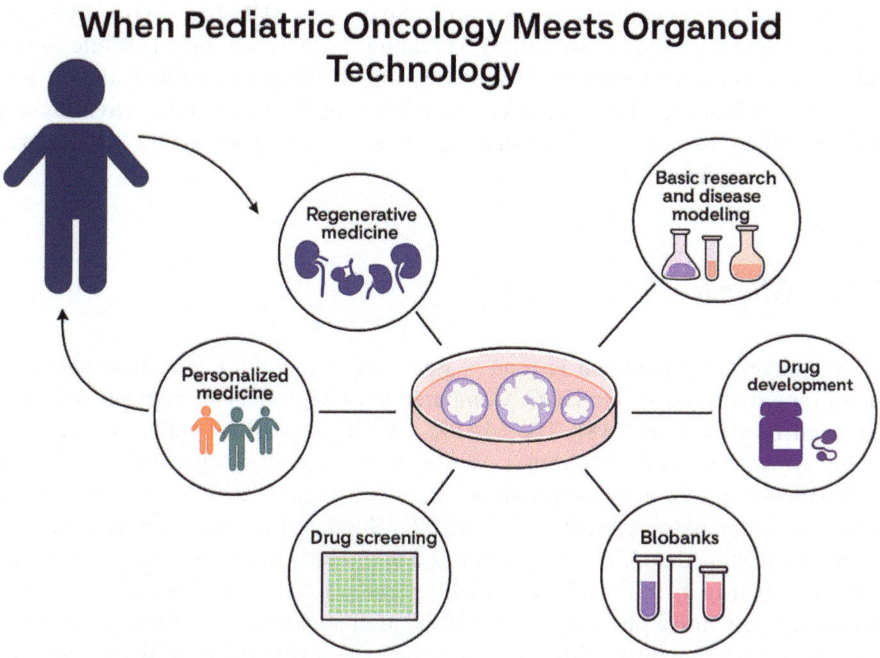

Fig. 7.1 Potential applications of organoids in pediatric cancer research. PDOs can be used for different biomedical applications such as basic research and disease modeling, drug screening and drug development, biobanking, personalized medicine, and regenerative medicine

availability of publications limited to a subset of the different tumor entities such as medulloblastoma (MB) and high-grade gliomas (HGGs) [10, 13].

7.3.1 Medulloblastoma

Medulloblastoma is the most frequent malignant brain tumor that mainly occurs in children [14]. MB has been classified into four molecular subgroups each associated with varying clinical risks: WNT, SHH, Group 3 (G3), and Group 4 (G4). The WNT and SHH subtypes generally have a favorable and intermediate prognosis. G3 and G4 are the most aggressive, often presenting metastases at diagnosis, and have the worst prognosis [14–16]. So far, organoids have only been used in a small number of studies to mimic MB. In the most recent study, in 2022, Jacob and colleagues published the development of organoids generated directly from tumors surgically resected from patients without cell dissociation. Based on a modified protocol derived from the Jacob study group [17], this approach is highlighted for its ability to preserve the integrity of tumor stroma, blood vessels, and immune infiltrate. They generated three viable organoids with heterogeneous morphology, intact blood vessels (CD31+), and molecular features (Sox2+) in SHH-MB. Also, drug and radiotherapy were tested in this study, revealing a synergistic effect. While organoids hold promise for modeling MB, addressing challenges in limited studies, reliance on established cell lines, unclear replication of the tumor microenvironment, and scalability issues are crucial to maximize their utility in advancing MB research and clinical management.

7.3.2 *Glioma*

Glioma covers all malignant lesions of glial origin and is a highly heterogeneous tumor ranging from LGG to HGG. Compared to LGG, HGGs are rare but also characterized by recurrence [18]. The MAPK, EGFR, and VEGF pathways are commonly affected in pediatric gliomas. While organoid models for adult glioblastoma (GBM) have seen significant progress, there is a notable lack of models tailored explicitly for pediatric patients [19]. In 2022, Sundar et al. successfully generated organoids from pediatric patients diagnosed with HGG involving embedded single cells into Matrigel, followed by a shaking culture. The study revealed different organoid proliferative phenotypes before and after standard care treatment (temozolimide and radiotherapy), utilizing immunohistochemistry microarrays for comprehensive evaluations. The results showed a prominent contrast in the response patterns, as the organoids exhibited resistance to the therapeutic regimen, while the glioma spheroids maintained sensitivity [20]. Recently, a protocol has been published outlining the generation of cerebellum and forebrain organoids from human-induced pluripotent stem cells (hiPSCs) [21]. The protocol acknowledges a workflow for modifying these organoids genetically by overexpressing genes, leading to the eventual production of tumor organoids. The authors detailed how to use MB and HGG organoids to assess their tumorigenic potential both in vitro and in vivo, which is promising for future organoid research.

7.4 Neuroblastoma

Neuroblastoma (NB) is the most common extracranial solid tumor in children, accounting for 7–10% of childhood malignancies and 15% of cancer-related mortality in patients less than 15 years old [22]. NB is primarily derived from precursor cells of the sympathetic nervous system, which frequently arise in the adrenal gland but can also develop in the neck and pelvis. Metastases are found in a majority of cases at diagnosis and are mainly localized in the bone marrow (BM) [23, 24]. This tumor is characterized by substantial genetic, morphological, and clinical heterogeneity [25]. Based on age, biological factors such as Shimada histopathology, DNA index, and MYCN amplification, patients are stratified into four risk groups: very-low, low, intermediate, or high-risk (HR) [26]. The low-risk group has a survival rate of >90% with surgery alone or spontaneous regression; the intermediate-risk group has a survival rate of >90% with surgery and chemotherapy; and the high-risk group has survival rates of 30–40% despite multimodality therapy (chemotherapy, radiation therapy, and stem cell transplantation) [22]. The MYCN gene encodes the NMYC protein, a transcription factor capable of triggering genes associated with self-renewal, pluripotency, metastasis, and angiogenesis. Additionally, it suppresses the expression of genes that promote differentiation, cell cycle arrest, and immune surveillance. Even if therapeutic advances in NB continue, they still remain a complex medical challenge, especially in HR cases [23]. The first patient-derived organoids (PDOs) model of NB was developed by Fusco and colleagues in 2019, who established six independent NB organoids from primary tumor derived from 4 HR-NB patients, stage M. PDOs were maintained viable in culture up to 2 months, the architecture was well recognized and mimicked the NB morphology. About the genomic profile, organoids and corresponding parental tumors showed high concordance exhibiting MYCB NB-specific chromosomal aberrations (MYCN amplification, 1p deletion, 11q loss). Authors revealed that PDOs retain stemness properties and heterogeneous cellular composition. Organoids are crucial in NB research since they are a representative preclinical in vitro tool that accurately reproduces the architecture, heterogeneity, and complicated biological processes of tumors of origin. This can overcome limitations associated with conventional cell and animal models for pharmacological testing [27].

7.5 Sarcomas

Sarcomas are a heterogeneous group of connective tissue malignancies of mesenchymal origin, including more than 70 subtypes [28]. Sarcomas can be divided into soft tissue sarcomas (STS) and Primary bone sarcomas (PBS) [29]. RMS and non-RMS STS represent the two major histological classes of STS, whereas osteosarcoma and Ewing sarcoma (ES) are PBS's most common histological subtypes [30]. STS typically arise in the extremities or retroperitoneum. The incidence of STS among all children who are younger than 20 years diagnosed with cancer is approximately 7–8, while representing only 1–2% of all cancer diagnoses in adults [31]. The 5-year survival rate for individuals diagnosed with soft tissue sarcomas decreases to 15% in cases where there is distant metastasis. Despite most bone sarcomas being detected at an early stage, the 5-year survival rate for patients diagnosed with distant stage is around 30% [32]. Sarcoma organoid generation is still relatively young and evolving, necessitating continued research and development.

7.5.1 Rhabdomyosarcoma

RMS is a mesenchyme-derived tumor and the most common childhood STS [33]. Based on histological features, two main subclasses have been defined in the pediatric population, namely, the embryonal (ERMS; 70% of all RMS) and the alveolar (ARMS; 20% of all RMS) subclass [34]. The PAX3/7-FOXO1 chromosomal translocation is associated with 85% ARMS [35]. Their 5-year survival rate ranges from 60 to 80% for patients with localized tumors but decreases to 20% for those who relapsed or had metastases at diagnosis. Innovative therapeutic strategies are even more required for patients with poor prognosis [36]. In 2023, Savary et al. developed 3D organoid model derived from relapsed pediatric fusion negative RMS. This model preserved the histological and molecular characteristics of aggressive fusion negative RMS tumors of derivation. It also preserves intra-tumoral heterogeneity even after several passages and cryopreservation as 3D cultures. This model was helpful in designing and evaluating a new drug combination [37].

7.5.2 Non-rhabdomyosarcoma Soft Tissue Sarcoma

Non-rhabdomyosarcoma soft tissue sarcomas (NRSTS) represent a small fraction, approximately 4%, of childhood cancers. This category encompasses a diverse array of mesenchymal extra skeletal malignancies. Due to their rarity, heterogeneity, and aggressive nature, managing NRSTS in children and adolescents poses significant complexities and challenges, and the number of well-established cell models is extremely limited [31]. Boulay et al. generate PTOs from synovial sarcoma, to explore chromatin-remodeling mechanisms. They conducted comprehensive epigenomic profiling across the genome [38]. Maloney et al. established the development of PTOs from skin fibrosarcoma, offering a novel platform for drug testing. They investigated the effects of the tyrosine kinase inhibitor imatinib and the anthracycline chemotherapy agent doxorubicin on these organoids [39].

7.5.3 Osteosarcoma

Osteosarcoma (OS) is the most common PBS, often diagnosed in children and teens [40]. An optimized treatment strategy for OS patients closely relies on the tumor histology, tumor size/grade, and the occurrence of the metastasis [41]. The 5-year survival rate of disseminated OS is relatively low, with a rate of 27%, compared to 60% of those without metastasis [42]. OS often spreads to lungs; around half of OS patients with lung metastasis experience recurrence even after surgery and combination therapy [43]. A few studies have investigated the use of organoids for disease modeling and drug sensitivity testing in OS. In 2020, He et al. reported an organoid culture system for lung metastatic OS tissue for the first time. They developed a protocol which led to maintaining and serially propagating the organoids for at least 6 months [44]. Johansson created OS PDOs, organoids displaying rounded structure in microscopy and secreting vascular endothelial growth factor (VEGF) in culture [45]. Subramaniam et al. generated multicell-type lung organoid models with OS cells and reported a significant reduction in OS cell growth after treatment with pimozide [46]. Nie et al. in 2022 established a panel of OS-derived organoids that maintained the expression patterns of the OS biomarkers (SOX9 and vimentin) and the expression patterns of the original tumor. The organoids showed active growth for months. They observed that anti-GPC3 immunotherapy can effectively suppress the growth of organoids [47].

7.5.4 Ewing Sarcoma

ES is an aggressive bone malignancy of adolescents and young adults and the second most common pediatric tumor, with higher incidence and poor prognosis [48]. In the last decades, local treatment that includes surgery/radiation and extensive use of multiagent chemotherapy (Vincristine, Doxorubicin, Cyclophosphamide/Ifosfamide, Etoposide) has improved the 5-year survival rate [49]. However, patients with an initial diagnosis of metastatic disease still have a very poor prognosis, while 25% of the patients with confined illness develop metastasis (mostly in lung and bone) within 2 years [50]. Maurer et al. developed ES organoids and monolayers from a metastatic pulmonary lesion from a patient with an inherited *BRCA1*-Associated RING Domain 1 (BARD1) mutation. The organoids demonstrated high sensitivity to poly (ADP-ribose) polymerase (PARP) inhibitors [51]. The same study group published the results of their second study, stating that the loss of BARD1 increases ES sensitivity to DNA damage, and that Guanylate-binding protein 1 (GBP1) expression contributes to DNA damage response in ES organoids [52]. Komatsu et al. were the first to use PD cell lines of CIC-DUX4 sarcoma to generate Ewing-like small round cell sarcoma organoids. Notably, drug sensitivity assays revealed a dose-dependent decrease in organoid size after treatment with two different concentrations of gemcitabine [53].

7.6 Retinoblastoma

Retinoblastoma (RB) is a tumor originating in the retina and is the most frequent eye cancer in children [54]. Presently, numerous organoid models have been created to investigate RB. In 2023, Srimongkol and colleagues observed that RB recurrence after chemo-reduction is common and is often managed with local (intra-arterial/intravitreal) chemotherapy. Using RB PDO, the authors screened with FDA-approved drugs, and candidate drugs were selected based on their cytotoxicity and potency [55]. Zheng and colleagues utilized a cell model derived from human embryonic stem cells (hESCs). This research's primary objective was to investigate the pathophysiological role of RB1 during human retinal development, which had not been extensively explored in previous studies. The researchers generated retinal organoids from RB1-null hESCs, created through CRISPR/Cas9 technology. Their findings revealed profuse expression of RB in retinal progenitor cells within the retinal organoids. Loss of RB1, however, led to increased entry into the S-phase and widespread apoptosis, resulting in reduced numbers of photoreceptors, ganglion cells, and bipolar cells. Interestingly, the mutation of RB1 in retinal organoids did not induce the formation of RB either in vitro or in the vitreous body of immunodeficient NOD/SCID mice. This work identifies a crucial function for RB1 in human retinal development and suggests that the deletion of RB1 alone is insufficient for tumor development, at least in human retinal organoids [56]. The broad spectrum of organoid models employed, ranging from RB tumor cells to induced pluripotent stem cells (iPSCs) and hESCs, highlights organoids' adaptability in RB research. These models have led robust investigations into drug responses and potential therapeutic agents with high efficacy.

7.7 Pediatric Kidney Tumors

Pediatric kidney tumors represent around 5% of pediatric tumors and can differ in many aspects, including cell of origin, genetics, pathology, and prognosis. The most common subtype is Wilms tumor (WT), followed by clear cell sarcoma of the kidney (CCSK), malignant rhabdoid tumor of the kidney (MRTK), renal cell carcinoma (RCC), and congenital mesoblastic nephroma (CMN) [57]. Although WT cell lines have contributed highly to the understanding of tumor biology, they still do not reflect the heterogeneous nature of WT. Cell models for other pediatric kidney cancer subtypes are rare. A model for WT was developed, employing a modified two-step protocol. Given the frequent association of WT with homozygous loss of the tumor suppressor WT1, the authors established a tumorigenesis model using human kidney organoids through the inducible deletion of WT1. The genetic knockout of WT1, implemented at various stages of renal differentiation, resulted in excessive proliferation of nephron progenitor cells (NPCs) at the expense of tubular and glomerular differentiation. A transcriptomic-wide analysis of the organoids demonstrated that the loss of WT1 recapitulates the molecular characteristics observed in a specific subset of WT patients [58]. Overall, while drug screens conducted on PDO have discovered patient-specific drug sensitivities and enhanced therapeutic strategies for WT, it is essential to report that there are certain limitations. PDOs may not fully replicate the tumor microenvironment or interactions with the immune system, potentially impacting drug responses observed in vitro versus in vivo. As organoid technology advances and the understanding of the molecular complexities of pediatric kidney tumors improves, there is significant potential for the development of tailored and efficient therapeutic interventions.

7.8 Pediatric Liver Tumors

Hepatoblastomas (HBs) and pediatric hepatocellular carcinomas (HCCs) cause approximately 80% of primary malignant liver tumors in children and adolescents. HBs are more common than HCCs but, from a prognostic point of view, the prognosis for pediatric HCC is dismal, with a 5-year event-free survival of <30%, in apparent contrast to HBs > 80% [59, 60]. Few organoid models have been created to investigate HBs, but no protocols have been described for growing HCC PDOs models [61]. As there are ongoing studies, HB organoids hold promising application prospects for future research endeavors.

7.9 The Future and Concluding Remarks

Organoid research represents a radical approach with profound implications for pediatric oncology [10]. Pediatric tumors display considerable inter- and intra-tumoral heterogeneity, necessitating personalized therapeutic strategies for optimal outcomes. Moreover, organoid-based drug screening platforms facilitate the rapid evaluation of drug efficacy and toxicity. Researchers can identify promising treatment candidates by screening various chemotherapeutic agents, targeted therapies, or experimental drugs against tumor organoids and expedite their translation into clinical trials. Cancer organoids that retain the genetic diversity of the primary tumor aim to identify the optimal treatment for an individual patient and indicate the potential development of drug resistance [62]. This awareness can guide further research to develop combination therapies and targeted strategies to address drug resistance. Another interesting application under development is the establishment of PDOs from circulating tumor cells (CTCs), which could allow tumor cell isolation from the blood of cancer patients and the subsequent establishment of PDOs and drug screening without invasive sampling. For instance, CTC-derived organoids have been established from lung cancer and prostate cancer cells [63]. This methodology could represent a breakthrough in pediatric cancer treatment, allowing earlier diagnosis and treatment response monitoring, while reducing the impact of invasive procedures on pediatric patients.

References

1. Waldron W, Altekruse SF, Kosary CL, Krapcho M, Neyman N, Aminou R, et al. SEER cancer statistics review 1975–2007. Bethesda, MD: National Cancer Institute; 2010.
2. Steliarova-Foucher E, Stiller C, Lacour B, Kaatsch P. International classification of childhood cancer, third edition. Cancer. 2005;103:1457–67. https://doi.org/10.1002/cncr.20910.

3. Gröbner SN, Worst BC, Weischenfeldt J, Buchhalter I, Kleinheinz K, Rudneva VA, et al. The landscape of genomic alterations across childhood cancers. Nature. 2018;555:321–7. https://doi.org/10.1038/nature25480.
4. Sweet-Cordero EA, Biegel JA. The genomic landscape of pediatric cancers:Implications for diagnosis and treatment. Science. 2019;363:1170–5. https://doi.org/10.1126/science.aaw3535.
5. Gatta G, Zigon G, Capocaccia R, Coebergh JW, Desandes E, Kaatsch P, et al. Survival of European children and young adults with cancer diagnosed 1995–2002. Eur J Cancer. 2009;45:992–1005. https://doi.org/10.1016/j.ejca.2008.11.042.
6. Saletta F, Seng MS, Lau LMS. Advances in paediatric cancer treatment. Transl Pediatr. 2014;3:156–82. https://doi.org/10.3978/j.issn.2224-4336.2014.02.01.
7. Landier W, Armenian S, Bhatia S. Late effects of childhood cancer and its treatment. Pediatr Clin N Am. 2015;62:275–300. https://doi.org/10.1016/j.pcl.2014.09.017.
8. Barbet V, Broutier L. Future match making: when pediatric oncology meets organoid technology. Front Cell Dev Biol. 2021;9:674219. https://doi.org/10.3389/fcell.2021.674219.
9. Lancaster MA, Knoblich JA. Generation of cerebral organoids from human pluripotent stem cells. Nat Protoc. 2014;9:2329–40. https://doi.org/10.1038/nprot.2014.158.
10. Lampis S, Galardi A, Di Paolo V, Di Giannatale A. Organoids as a new approach for improving pediatric cancer research. Front Oncol. 2024;14:1414311. https://doi.org/10.3389/fonc.2024.1414311.
11. Merchant TE, Pollack IF, Loeffler JS. Brain tumors across the age spectrum: biology, therapy, and late effects. Semin Radiat Oncol. 2010;20:58–66. https://doi.org/10.1016/j.semradonc.2009.09.005.
12. Formentin C, Joaquim AF, Ghizoni E. Posterior fossa tumors in children: current insights. Eur J Pediatr. 2023;182:4833–50. https://doi.org/10.1007/s00431-023-05189-5.
13. Riedel NC, De Faria FW, Alfert A, Bruder JM, Kerl K. Three-dimensional cell culture systems in pediatric and adult brain tumor precision medicine. Cancer. 2022;14:5972. https://doi.org/10.3390/cancers14235972.
14. Louis DN, Perry A, Wesseling P, Brat DJ, Cree IA, Figarella-Branger D, et al. The 2021 WHO classification of tumors of the central nervous system: a summary. Neuro-Oncology. 2021;23:1231–51. https://doi.org/10.1093/neuonc/noab106.
15. Duan J, Wang Y. Modeling nervous system tumors with human stem cells and organoids. Cell Regen. 2023;12:4. https://doi.org/10.1186/s13619-022-00150-7.
16. Cavalli FMG, Remke M, Rampasek L, Peacock J, Shih DJH, Luu B, et al. Intertumoral heterogeneity within medulloblastoma subgroups. Cancer Cell. 2017;31:737–54.e6. https://doi.org/10.1016/j.ccell.2017.05.005.
17. Jacob F, Salinas RD, Zhang DY, Nguyen PTT, Schnoll JG, Wong SZH, et al. A patient-derived glioblastoma organoid model and biobank recapitulates inter- and intra-tumoral heterogeneity. Cell. 2020;180:188–204.e22. https://doi.org/10.1016/j.cell.2019.11.036.
18. Thorbinson C, Kilday JP. Childhood malignant brain tumors: balancing the bench and bedside. Cancer. 2021;13:6099. https://doi.org/10.3390/cancers13236099.
19. Grigore FN, Yang SJ, Chen CC, Koga T. Pioneering models of pediatric brain tumors. Neoplasia. 2023;36:100859. https://doi.org/10.1016/j.neo.2022.100859.
20. Sundar SJ, Shakya S, Barnett A, Wallace LC, Jeon H, Sloan A, et al. Three dimensional organoid culture unveils resistance to clinical therapies in adult and pediatric glioblastoma. Transl Oncol. 2022;15:101251. https://doi.org/10.1016/j.tranon.2021.101251.
21. Lago C, Gianesello M, Santomaso L, Leva G, Ballabio C, Anderle M, et al. Medulloblastoma and high-grade glioma organoids for drug screening, lineage tracing, co-culture and in vivo assay. Nat Protoc. 2023;18:2143–80. https://doi.org/10.1038/s41596-023-00839-2.
22. Ishola TA, Chung DH. Neuroblastoma. Surg Oncol. 2007;16:149–56. https://doi.org/10.1016/j.suronc.2007.09.005.
23. Louis CU, Shohet JM. Neuroblastoma: molecular pathogenesis and therapy. Annu Rev Med. 2015;66:49–63. https://doi.org/10.1146/annurev-med-011514-023121.

24. Sharp SE, Gelfand MJ, Shulkin BL. Pediatrics: diagnosis of neuroblastoma. Semin Nucl Med. 2011;41:345–53. https://doi.org/10.1053/j.semnuclmed.2011.05.001.
25. Esposito MR, Aveic S, Seydel A, Tonini GP. Neuroblastoma treatment in the post-genomic era. J Biomed Sci. 2017;24:14. https://doi.org/10.1186/s12929-017-0319-y.
26. Matthay KK, Maris JM, Schleiermacher G, Nakagawara A, Mackall CL, Diller L, et al. Neuroblastoma. Nat Rev Dis Primer. 2016;2:16078. https://doi.org/10.1038/nrdp.2016.78.
27. Fusco P, Parisatto B, Rampazzo E, Persano L, Frasson C, Di Meglio A, et al. Patient-derived organoids (PDOs) as a novel in vitro model for neuroblastoma tumours. BMC Cancer. 2019;19:970. https://doi.org/10.1186/s12885-019-6149-4.
28. Fletcher CDM, Bridge JA, Hogendoorn PCW, Mertens F. WHO classification of tumours of soft tissue and bone. Lyon: IARC Press; 2013. Available at: https://publications.iarc.fr/Book-And-Report-Series/Who-Classification-Of-Tumours/WHO-Classification-Of-Tumours-Of-Soft-Tissue-And-Bone-2013.
29. Gounder MM, Agaram NP, Trabucco SE, Robinson V, Ferraro RA, Millis SZ, et al. Clinical genomic profiling in the management of patients with soft tissue and bone sarcoma. Nat Commun. 2022;13:3406. https://doi.org/10.1038/s41467-022-30496-0.
30. Lupo PJ, Spector LG, O'Brien S, Schiffman JD, Hettmer S. Epidemiology of bone and soft tissue sarcomas. In: Pediatric oncology. Springer Science and Business Media Deutschland GmbH; 2020. p. 1–16.
31. Ferrari A, Sultan I, Huang TT, Galindo CR, Shehadeh A, Meazza C, et al. Soft tissue sarcoma across the age spectrum: a population-based study from the surveillance epidemiology and end results database. Pediatr Blood Cancer. 2011;57:943–9. https://doi.org/10.1002/pbc.23252.
32. Longhi A, Errani C, De Paolis M, Mercuri M, Bacci G. Primary bone osteosarcoma in the pediatric age: state of the art. Cancer Treat Rev. 2006;32:423–36. https://doi.org/10.1016/j.ctrv.2006.05.005.
33. Li J, Thompson TD, Miller JW, Pollack LA, Stewart SL. Cancer incidence among children and adolescents in the United States, 2001–2003. Pediatrics. 2008;121:e1470–7. https://doi.org/10.1542/peds.2007-2964.
34. Skapek SX, Ferrari A, Gupta A, Lupo PJ, Butler E, Shipley J, et al. Rhabdomyosarcoma. Nat Rev Dis Primer. 2019;5:1. https://doi.org/10.1038/s41572-018-0051-2.
35. Parham DM, Barr FG. Classification of rhabdomyosarcoma and its molecular basis. Adv Anat Pathol. 2013;20:387–97. https://doi.org/10.1097/PAP.0b013e3182a92d0d.
36. Arndt CAS, Bisogno G, Koscielniak E. Fifty years of rhabdomyosarcoma studies on both sides of the pond and lessons learned. Cancer Treat Rev. 2018;68:94–101. https://doi.org/10.1016/j.ctrv.2018.06.013.
37. Savary C, Luciana L, HuchedéP TA, Coquet C, Broustal M, et al. Fusion negative rhabdomyosarcoma 3D organoids to predict effective drug combinations: a proof-of-concept on cell death inducers. Cell Rep Med. 2023;4:101339. https://doi.org/10.1016/j.xcrm.2023.101339.
38. Boulay G, Cironi L, Garcia SP, Rengarajan S, Xing YH, Lee L, et al. The chromatin landscape of primary synovial sarcoma organoids is linked to specific epigenetic mechanisms and dependencies. Life Sci Alliance. 2021;4:e202000808. https://doi.org/10.26508/lsa.202000808.
39. Maloney E, Clark C, Sivakumar H, Yoo K, Aleman J, Rajan SAP, et al. Immersion bioprinting of tumor organoids in multi-well plates for increasing chemotherapy screening throughput. Micromachines. 2020;11:208. https://doi.org/10.3390/mi11020208.
40. Rojas GA, Hubbard AK, Diessner BJ, Ribeiro KB, Spector LG. International trends in incidence of osteosarcoma (1988–2012). Int J Cancer. 2021;149:1044–53. https://doi.org/10.1002/ijc.33673.
41. Gill J, Ahluwalia MK, Geller D, Gorlick R. New targets and approaches in osteosarcoma. Pharmacol Ther. 2013;137:89–99. https://doi.org/10.1016/j.pharmthera.2012.09.003.
42. Zhang Y, Yang J, Zhao N, Wang C, Kamar S, Zhou Y, et al. Progress in the chemotherapeutic treatment of osteosarcoma. Oncol Lett. 2018;16:6228–37. https://doi.org/10.3892/ol.2018.9434.

43. Nataraj V, Rastogi S, Khan SA, Sharma MC, Agarwala S, Vishnubhatla S, et al. Prognosticating metastatic osteosarcoma treated with uniform chemotherapy protocol without high dose methotrexate and delayed metastasectomy: a single center experience of 102 patients. Clin Transl Oncol. 2016;18:937–44. https://doi.org/10.1007/s12094-015-1467-8.
44. He A, Huang Y, Cheng W, Zhang D, He W, Bai Y, et al. Organoid culture system for patient-derived lung metastatic osteosarcoma. Med Oncol. 2020;37:105. https://doi.org/10.1007/s12032-020-01429-y.
45. Johansson S. Patient-derived organoid culture for 3D culture of colorectal cancer, renal cancer and osteosarcoma [Internet]. Dissertation. 2019. Available at: https://urn.kb.se/resolve?urn=urn:nbn:se:uu:diva-391043.
46. Subramaniam D, Angulo P, Ponnurangam S, Dandawate P, Ramamoorthy P, Srinivasan P, et al. Suppressing STAT5 signaling affects osteosarcoma growth and stemness. Cell Death Dis. 2020;11:149. https://doi.org/10.1038/s41419-020-2335-1.
47. Nie J, Yang T, Li H, Li S, Li T, Ye H, et al. Frequently expressed glypican-3 as a promising novel therapeutic target for osteosarcomas. Cancer Sci. 2022;113:3618–32. https://doi.org/10.1111/cas.15521.
48. Grünewald TGP, Cidre-Aranaz F, Surdez D, Tomazou EM, de Álava E, Kovar H, et al. Ewing sarcoma. Nat Rev Dis Primer. 2018;4:1–22. https://doi.org/10.1038/s41572-018-0003-x.
49. Balamuth NJ, Womer RB. Ewing's sarcoma. Lancet Oncol. 2010;11:184–92. https://doi.org/10.1016/S1470-2045(09)70286-4.
50. Riggi N, Suvà ML, Stamenkovic I. Ewing's sarcoma. N Engl J Med. 2021;384:154–64. https://doi.org/10.1056/NEJMra2028910.
51. Maurer LM, Venier RM, Julian C, Bailey KM. Abstract A14: exploiting DNA damage repair defects to enhance PD-L1 expression in Ewing sarcoma. Cancer Res. 2020;80:A14. https://doi.org/10.1158/1538-7445.PEDCA19-A14.
52. Maurer LM, Daley JD, Mukherjee E, Venier RE, Julian CM, Bailey NG, et al. BRCA1-associated RING domain-1 (BARD1) loss and GBP1 expression enhance sensitivity to DNA damage in Ewing sarcoma. Cancer Res Commun. 2022;2:220–32. https://doi.org/10.1158/2767-9764.CRC-21-0047.
53. Komatsu A, Matsumoto K, Yoshimatsu Y, Sin Y, Kubota A, Saito T, et al. The CAM model for CIC-DUX4 sarcoma and its potential use for precision medicine. Cells. 2021;10:2613. https://doi.org/10.3390/cells10102613.
54. Villegas VM, Hess DJ, Wildner A, Gold AS, Murray TG. Retinoblastoma. Curr Opin Ophthalmol. 2013;24:581–8. https://doi.org/10.1097/ICU.0000000000000002.
55. Srimongkol A, Laosillapacharoen N, Saengwimol D, Chaitankar V, Rojanaporn D, Thanomchard T, et al. Sunitinib efficacy with minimal toxicity in patient-derived retinoblastoma organoids. J Exp Clin Cancer Res. 2023;42:39. https://doi.org/10.1186/s13046-02302608-1.
56. Zheng C, Schneider JW, Hsieh J. Role of RB1 in human embryonic stem cellderived retinal organoids. Dev Biol. 2020;462:197–207. https://doi.org/10.1016/j.ydbio.2020.03.011.
57. Ooms AHAG, Calandrini C, De Krijger RR, Drost J. Organoid models of childhood kidney tumours. Nat Rev Urol. 2020;17:311–3. https://doi.org/10.1038/s41585-020-0315-y.
58. Waehle V, Ungricht R, Hoppe PS, Betschinger J. The tumor suppressor WT1 drives progenitor cell progression and epithelialization to prevent Wilms tumorigenesis in human kidney organoids. Stem Cell Rep. 2021;16:2107–17. https://doi.org/10.1016/j.stemcr.2021.07.023.
59. Bhagat P, Vij M, Raju LP, Gowrishankar G, Menon J, Shanmugam N, et al. Update on the pathology of pediatric liver tumors: a pictorial review. Diagnostics. 2023;13:3524. https://doi.org/10.3390/diagnostics13233524.
60. Ranganathan S, Lopez-Terrada D, Alaggio R. Hepatoblastoma and pediatric hepatocellular carcinoma: an update. Pediatr Dev Pathol. 2020;23:79–95. https://doi.org/10.1177/1093526619875228.

61. Digiacomo G, Serra RP, Turrini E, Tiri A, Cavazzoni A, Alfieri R, et al. State of the art and perspectives in pediatric hepatocellular carcinoma. Biochem Pharmacol. 2023;207:115373. https://doi.org/10.1016/j.bcp.2022.115373.
62. Bartfeld S, Clevers H. Stem cell-derived organoids and their application for medical research and patient treatment. J Mol Med. 2017;95:729–38. https://doi.org/10.1007/s00109-017-1531-7.
63. Gao F, Zhang P, Zhou C, Li J, Wang Q, Zhu F, et al. Frequent loss of PDCD4 expression in human glioma: possible role in the tumorigenesis of glioma. Oncol Rep. 2007;17:123–8.

Chapter 8
Engineering Transplantable Organoids for Children with Pediatric Disorders: Organoids in Pediatric Surgery

F. Tansu Salman

Abstract Organoid technology is rapidly transforming the landscape of pediatric surgery by offering unprecedented opportunities in regenerative medicine, patient-specific modeling, and surgical planning. Derived from pluripotent or tissue-specific stem cells, organoids replicate key structural and functional aspects of developing organs, making them particularly valuable in the context of pediatric and fetal biology. In regenerative medicine, organoids provide a foundation for creating tissue-engineered grafts tailored for congenital anomalies such as intestinal atresia, biliary malformations, or bladder exstrophy. Fetal organoids, which mimic early-stage organ development, enable researchers to investigate the pathogenesis of congenital conditions and optimize the timing and strategy for in utero or early-life interventions. In pediatric oncology, tumor-derived organoids offer platforms for studying rare childhood cancers and testing targeted therapies in a patient-specific manner, advancing the goals of precision medicine. Furthermore, organoids are beginning to support personalized surgical planning, by modeling the anatomical and functional variations in diseased tissues prior to complex operations. As the field progresses, future pediatric surgery research will likely focus on integrating organoid-based diagnostics, drug screening, and bio-fabrication technologies to develop living implants and optimize post-operative outcomes. Ultimately, organoids hold the potential to revolutionize pediatric surgical care by bridging developmental biology with clinical innovation in a child-centered framework.

Keywords Pediatric surgery oriented organoid research · Transplantable organoids · Fetal organoids · Patient-specific modeling

F. Tansu Salman (✉)
Bahçeşehir University, Faculty of Medicine, Department of Pediatric Surgery, Istanbul, Turkey

O. S. Somuncu (ed.), *Organoids for Preclinical Disease Modeling and Personalized Medicine*, Stem Cell Biology and Regenerative Medicine 78,
https://doi.org/10.1007/978-3-032-10024-5_8

8.1 Introduction

Organoids are intriguing scientists widely as they will provide great benefits in the field of medicine and in all areas related to health. Organoids are a miniaturized, simplified three-dimensional version of an organ that can mimic the basic functional, structural, and biological complexity of the original organ. These newly produced miniature organs are derived in vitro from embryonic stem cells or induced pluripotent stem cells. As a result of their self-renewal and differentiation capacities, these miniature, lab-grown, three-dimensional formations can organize themselves, mimic the architecture and function of real organs, and are applicable in various areas of pediatric medicine [1].

Before examining organoids, it may be useful to write about regenerative medicine, in other words tissue engineering. Tissue engineering, first described by Langer and Vacanti, is an interdisciplinary field that applies the principles of engineering and life [2]. The loss or failure of an organ or tissue is one of the most frequent, devastating, and costly problems in human health care especially in the field or pediatric surgery. For this problem, tissue engineering is a progressing new area that seems to be a solution.

When I was working at the University of California at Los Angeles, a study was carried out for the purpose of investigate if fetal and newborn allografts can survive in an immunocompetent host? To find an answer to this question at the surgical laboratory of the Pediatric Surgery department, allografts of renal or adrenal tissue (1 mm^2) were implanted under the renal capsule of the recipient animals. When 44 fetal grafts were implanted into newborn recipients, there was a modest increase in size of 7.1 ± 7.5 fold when the grafts were assessed between 8 and 10 days after implantation. It was discussed and assumed that selected fetal tissue can grow and mature in the adult recipient [3].

At that time and following years, curiosity and studies on this subject were increasing and scientists were starting to plan many studies on tissue transplantation. Now we know that tissue engineering is an interdisciplinary field that applies the principles of organoids towards the development of biological substitutes that restore, maintain, or improve tissue function. So, organoids can generate functional tissues that could be transplanted to repair or replace damaged organs in children with non-functioning liver, lung, or kidneys and even organs with a hollow lumen such as esophagus, intestine, urinary tract.

Although clinical studies are not much started yet, promising experimental studies in vitro are very popular now a days. As seen in various scientific studies, decellularization is the primary technique which comprises the elimination of cellular apparatuses from the tissue to obtain biological scaffold substances by forming extracellular matrix and then recellularization is a progression of producing practical cells by seeding into the decellularized organ scaffold to generate entirely operative simulated organ [4].

These cell-seeded scaffolds are either cultured in vitro to synthesize tissues which can then be implanted into an injured site or are implanted directly into any site, using the body's own systems where regeneration of tissues or organs is induced in vivo. Researches are moving at a great pace in regenerative medicine, and attempts are being made to engineer organoids in vitro practically for every tissue and organ in the body [5].

Scientists are in a quest for better techniques for more durable scaffolds, more effective cells, and strong signals. Obtaining many different epithelial organoids from biopsies taken from various sites in pediatric patients will thus provide a field for improved medical education, modeling diseases, and developing new treatments especially for personalized medicine. Organoids also have a significant potential for diagnosing and treating congenital and acquired conditions in children and especially understanding the prenatal development to learn the true etiology of the defects. In this chapter, it is aimed to examine the use and research potential of organoids in the field of pediatric surgery with in the topics of regenerative medicine, pre-surgical planning and evaluation, modeling diseases, personalized medicine, fetal procedures, and childhood neoplasms.

8.2 Regenerative Medicine and Tissue Engineering

As mentioned above, organoids may be alternative to transplantation and non-biological tools as to regenerate damaged or distorted organs instead of replacing them. Anomalies of gastrointestinal system is ideal candidate for this process. Short bowel syndrome is a serious case which happens after extensive resections to ensure survival in some fatal and serious cases such as intestinal atresia, volvulus, gastroschisis, long segment Hirschsprung disease, and necrotizing enterocolitis. In short bowel syndrome, the main problem is intestinal failure that is the lack of sufficient functional intestine to sustain survival [6]. Some surgical procedures for lengthening the functional intestine and increasing the mucosal surface area are being performed; however, they are not usually sufficient. Intestinal transplantation in infants is also not successful because of technical difficulties and immunological problems. Kitano and his friends managed to generate intestinal grafts by recellularization native intestinal matrix with human-induced pluripotent stem cell-derived epithelium and human endothelium and showed nutrient absorption after transplantation in rats [7]. As studies on organoids progress, it become very nice to see that to construct autologous intestinal mucosal grafts by patient-derived organoids may be helpful. Laween and her friends from London showed in their detailed study that intestinal graft reconstruction using jejunal organoids can expand efficiently to provide intestinal mucosal function [1].

Long gap esophageal atresia is another very important anomaly in newborns. Although pediatric surgeons are quite skillful in primary repairing the esophagus, there may be some major postoperative complications like leakage and strictures. In these very serious cases, esophageal reconstruction may be necessary by performing gastric, jejunal, or colonic segments interpositions. These surgical procedures impose new complications and risks on the child. So, engineering a new esophagus for the child seems very reasonable. Studies for tissue engineering in long gap esophageal atresia by using decellularized material as scaffolds are very promising [8]. Urbani showed that tissue-engineered esophagus is much better than the esophageal substitutions. Combining decellularized scaffolds with patient-derived cells holds promise for regeneration of tissue defects. In their study on multi-stage bioengineering of a layered esophagus with in vitro expanded muscle and epithelial adult progenitors, they postulated that optimized two-stage approach produces a fully re-populated, structurally organized, and pre-vascularized esophageal substitute, which could become an alternative to current esophageal substitutes [9]. Long-term cryopreservation of decellularized esophagus for tissue engineering clinical application may also be possible [10].

Urethral tract problems may also need reconstructions by using regeneration and organoid procedures. For example, hypospadias is another congenital defect that forces the pediatric urologist. It is not a fatal condition, but it is a great burden to the child and his family. There is a defect on the ventral side of the penis and pediatric urologists try to make reconstruction by performing urethroplasty. Unfortunately, leakage can be seen quite often that construction of a new urethra may be useful to perform by tissue engineering [11]. In our molecular biology laboratory at Bahcesehir University School of Medicine, we made an in vitro study. In our study, waste saphenous veins were collected and decellularization process was applied. Optimized decellularization method that was modeled by our group was used with combination of physical and chemical methods [4]. We thought using saphenous vein is advantageous because it is a tubular structure and has a width which is close to that of a urethra and not difficult to obtain. The muscle layer is thought to create a more suitable extracellular matrix structure for urethra. Newborn circumcised prepuce stem cells were seeded onto biological scaffold. We seeded vascular smooth muscle cells in innermost layer, and under this layer, we seeded endothelial cells. For the outermost layer, we seeded foreskin stem cells isolated from the circumcised tissues [12]. Foreskin stem cells are advantageous because of their great proliferation rate and they have higher differentiation potential than other mesenchymal stem cell sources. It is also very easy to obtain for a pediatric surgeon and urologist since circumcision is a very common surgical procedure. If this method can be used at the clinics for reconstruction of the boy's urethra, we believe that tissue-engineered urethra would overcome complications associated with hypospadias. Huang managed to reconstruct penile urethra with the 3-dimensional porous bladder acellular matrix in a rabbit model [13]. Urethra repair by tissue engineering has been extensively studied in vitro and in laboratory animals but is not routinely used in clinical practice yet [14]. More improvements to pre-clinical study designs are required to improve clinical tissue engineering technologies.

Organoid bladder engineering especially for exstrophy vesica, neurogenic bladder, malignancies, and other urinary system problems could be possible. The place of organoids in the search for bladder cancer treatments will be discussed below [15]. For kidney, engineered rat kidney scaffolds by decellularization and recellularization method by human umbilical cord blood-derived endothelial cells studied and presented urine making accompanied by minor macromolecular sieving and reabsorption capacity. It was also shown to provide a useful insight into kidney organoid vascularization in vitro and presents a tool for further studies of kidney development and drug testing both for research purposes and pre-clinical applications [16, 17].

Respiratory defects in children are another major problem. Experimental animal studies on trachea are very promising. Especially trial efforts on tracheal scaffold strategies for pediatric transplantation in a rabbit model are very important guiding scientific steps. Hung performed segmental organ decellularization on trachea and employed autotransplantation on rabbit models [18]. The first successful clinically engineered tissues have been applied in a child with tracheal stenosis in 2010. Decellularized scaffolds was used to regenerate child's airway. Elliott et al. described the case of a child who received a stem-cell-based tracheal replacement and report findings after 2 years of follow-up. He had a functional airway and had returned to school [19]. Follow-up of the first pediatric, stem-cell-based, tissue-engineered transplant shows potential for this technology but highlights the need for further research.

Recellularized lung scaffold into rats was also performed with possible oxygen and carbon dioxide interchange [20, 21]. The scientists believe that physiological reengineering of the pulmonary vascular tree will significantly aid future clinical applications of bioengineered systems. Zhou, H. et al. studied bioengineering human lung grafts using porcine matrix. They concluded that engineering and transplantation of viable lung grafts based on lung scaffolds and human endothelial and epithelial cells is technically feasible [22]. Further graft maturation may be necessary to enable higher-level functions such as mucociliary clearance and ventilation-perfusion matching.

Bochdalek hernia, in other words congenital diaphragmatic hernia with respiratory distress at birth, is a very serious problem of newborn. The main problem in this anomaly is pulmonary insufficiency due to pulmonary hypoplasia and hypertension. Lung hypoplasia is thought to be caused by the compression of the lung from the herniated abdominal contents during the intrauterine period. To get good results in treatment, we must correct the lung disorder and seek remedies to prevent compression in intrauterine life. So, the defect at the diaphragm must be repaired before the hypoplasia happens. That is why lung organoids may be life saver in this catastrophic anomaly. It will also be very useful to make new experiments in utero to prevent pulmonary compression [24].

Although liver transplantation is now very common, easy, and routine operation, hepatobiliary anomalies in children are not always that easy to handle. Mazza demonstrates decellularization of human liver and repopulation with derived human liver cells. This is a key advance in bioartificial liver development [25]. The article by Wai et al. on liver organoids gives us very good information about the profiling to inform the success of operations and guiding the managements [26].

In molecular biology department at Bahcesehir University School of Medicine, an interesting study was held in which an artificial skin was engineered in vitro. This study is a reference investigation for skin engineering based on placental stem cells and biological scaffolds. Waste placental tissues were collected and decellularization process was applied. Decellularized and normal placental tissues were compared by immunohistochemistry. Donor's own placental stem cells were seeded onto biological scaffold and were differentiated into skin related cell types [12, 27]. In pediatric surgery, extra skin tissue was needed most of the time for treating various severe cases where skin was lost, for example, burns, trauma, severe infections, and after wide skin excisions of superficial tumors. Also, in some congenital defects, pediatric surgeon may need extra skin tissue to cover the defect such as omphalocele, exstrophy vesica, and meningomyelocele. There is also another interesting study about skin to assess whether perfusion-decellularization technology could be applied to facial models. Acellular scaffolds were produced from the porcine ear and concluded that it may be a valuable platform to treat facial deformities using regenerative medicine approaches [28].

Research is moving at a great pace in regenerative medicine in many fields. Hopkins et al. bioengineered heart valves to evaluate the outcome of replacing them by human and baboon heart valve scaffolds. For both mechanistic research and possible regulatory utility, these studies were undertaken to evaluate the suitability of the baboon as a subhuman primate model for testing clinical prototype. Bioengineered replacement valves from both species were hemodynamically equivalent to native valves. The tested bioengineering methods reduced antigenicity of both human and baboon valves [29].

Attempts are being made to engineer in vitro practically every tissue and organ in the body such as pancreas, bone and muscles, vessels, cornea, etc. Scientists are in quest for better techniques for more durable scaffolds, more effective cells, and strong signals.

8.3 Pre-surgical Planning, Evaluation, Modeling Congenital Disorders, or Acquired Diseases

Organoids enable testing the efficacy and safety of drugs before clinical application, which helps to minimize the need for invasive procedures. As we have seen so far, organoids can replicate the anatomy and even the pathology of specific organs. From the surgical point of view, pre-surgical modeling has a great advantage to surgeons for planning the complex cases. In addition, by testing interventions on organoids, surgeons can better anticipate surgical outcomes by evaluating the prognosis and potential complications.

The Kasai operation is a very good example for this issue. As mentioned in Wai's article above, liver organoids may predict the outcome of Kasai portoenterostomy. They studied the organoids generated from liver biopsies from biliary atresia patients at and after portoenterostomy. Their results are productive data to evaluate the timing of the operation, function of the ducts and the liver, and possible need of liver transplantation [26]. In this review study, it was addressed useful strategies to generate and use of organoids in human hepatobiliary disease modeling, followed by a discussion of their potential application in drug development and therapeutics, as well as their strengths and limitations [25, 26].

Liver organoids derived from patient biopsy tissues of patients with Alagille syndrome manifesting cholestatic jaundice could reflect the in vivo pathology. As in intestinal organoids, specific patient-derived liver organoids may allow the development of in vitro liver disease models such as modelling Alagille syndrome [30]. This may give the opportunity to study pathogenesis, identify the prognostic factors, and predict drug response in each biliary atresia patient and other liver diseases for patient-oriented medicine.

Gastric and intestinal organoids also may be very helpful for aiding surgeons for presurgical modeling and predicting outcomes [31]. Advances in organoid culture technology have provided a greater understanding of disease pathogenesis which enable modeling in vitro human development and various diseases [32]. Intestinal organoids have allowed the development of new in vitro models to study various gastrointestinal diseases expanding our understanding of the pathophysiology of diseases and promoting the development of innovative therapies [30]. Chusilp's team presented their outstanding new theories and studies in their articles consisting necrotizing enterocolitis modeling, organoids with enteric nervous system, liver organoids in biliary atresia, drug development, and gene therapy for cystic fibrosis disease [30].

Hirschsprung disease (HD) is a congenital colonic aganglionosis with abnormal enteric nervous system development. Many HD patients develop enterocolitis despite surgical resection. The pathophysiology of this inflammatory process is not understood clearly. Function of matched ganglionic and aganglionic regions was investigated utilizing organoids generated from these tissues. Organoids derived from ganglionic and aganglionic regions of HD patients were similar in epithelial cell differentiation. It is observed that transcriptional and functional changes in colonic epithelium are persevered in the ganglionic colon used for pull-through surgery. This may explain persistence of enterocolitis despite surgical excision of aganglionic colon and subsequent endorectal pull-through performed with ganglionic colon during correction of HD [33]. Because of a failure of enteric neural crest cells from up to caudal part of the gut, the ganglions are missing at the distal part of the intestines. Organoid-related experiments on HD successfully showed that human intestinal organoids contained Cajal cells (pacemaker cells that create bioelectrical waves that control smooth muscle contractions in colon) and consisted of nitric oxide synthase neurons (the important signaling molecule that results in smooth muscle relaxation along the intestines) [30]. These findings of intestinal organoids combined with enteric neural crest cells containing neurons, ganglions, Cajal and signaling molecules besides epithelium are very important to clarify the pathophysiology and prognosis of HD and other motility disorders. Combining tissue engineering and cellular replacement therapies represents a new strategy for treating enteric neuropathies, particularly HD.

Necrotizing enterocolitis (NEC) which is an acquired disease with a high mortality rate in preterm newborns is caused by bacterial invasion into the intestinal wall. It is characterized by inflammation and mucosal or even deeper necrosis of the intestinal wall. In recent years, scientists have explored the possibility of adapted individualized prevention and treatment of NEC [34]. Although the pathogenesis of this disease is not clearly known, gut immaturity, impaired gut barrier, microbiome, and feeding disorders may be the possible causes. Hypoxia and various manipulations at the intensive care unit such as ventilation problems lead to circulatory insufficiency and increased bacterial translocation that can lead to sepsis in the newborn [35].

Organoids derived from human pluripotent stem cells or from fetal intestine are great source for disease modeling that is to gain insights into intestinal development and NEC pathogenesis. These organoids show close resemble immature human intestinal epithelium in ability of maturation, digestive functions, and host-defense mechanisms. Mimicking intestinal injury of NEC will help us to understand the pathogenesis and prognosis of intestinal diseases of prematurity more clearly [30, 36].

Organoids may be helpful to researchers to model congenital abnormalities. Reviewing the literature, we meet valuable studies on pediatric gastric organoids that can be used as modelling diseases. McCracken and colleagues described a detailed protocol to generate 3-dimensional human intestinal tissue in vitro from human pluripotent stem cells. They managed to derive intestinal organoids for models of human development and diseases [37]. Another study is from Great Ormond Street, London, in which it was shown that gastric organoids can be derived reliably from gastric epithelium in children. They also studied on gastric fundus specifications. They achieved mucosal biopsies by gastroscopy or biopsy samples taken at the time of performing or closure of gastrostomy. In their paper, potential application of these patient-derived gastric organoids were emphasized: drug screening, diagnosis, gene editing and disease modelling, and even regenerative medicine. It was discussed that gastric or intestinal organoids may be used to improve mucosal healing in models of gastrointestinal inflammatory diseases [38, 39]. Gastric organoids from healthy adults also may be helpful to investigate gastric hormones, cancer, stomach diseases like gastritis, and of course *Helicobacter pylori* infections [40]. Any other infection, effective pathogens, antibiotic selection, and screening the therapies may also be evaluated in more detail and accurate by gastric and other organoids. The investigations of vaccinations using organoids are more speculative at present [38, 40].

8.4 Patient-Oriented Medicine

Organ specific adult stem cells organoids can also be used in personalized medicine by allowing ex vivo testing of drug responses as well as development of transplantation methodologies, modeling diseases, and pre-surgical planning. Advances in stem cell culturing and organoids open the door to preclinical personalized drug testing and make them a valuable tool with great potential to reduce or abolish animal tests that can only provide a limited prediction for the human population in general. It seems gratifying that organoids derived from a child's tissues may also be used to personalize treatment. Studies on lung organoids by Barkauskas summarized the different methods for generating organoids from cells isolated from human and mouse lungs and discussed the potential and limitations in organ biology and for developing new drugs for lung disorders such as cystic fibrosis and asthma [41]. So, organoids derived from a child's tissues can be used to personalize treatment and help assess drug responses specific to the patient's mutation profile.

Cystic fibrosis (CF) is the defects of exocrine glands that chloride transport is impaired and that results in abnormally viscous secretion. It is inherited as an autosomal recessive trait where there is a dysfunction in CF transmembrane conductance regulator proteins (CFTR). This pathology expressed largely in epithelial cells of airways, GI tract (gut, pancreas, biliary), sweat glands, and GU system (testes). Intestinal problems are serious complications in CF. The pediatric surgeons know that 80–90% of infants with meconium ileus have CF; however, respiratory problems are more severe and usually fatal. Modeling of CF by intestinal organoids provided the developments of modulating drugs. Patient-specific intestinal organoids cultured from rectal biopsy tissues of infants with CF-related meconium ileus were used for genetic studies and development of gene modulating therapy for CF [30, 42]. These innovations are very important mile stones in medical science to show the developments of drug responses in individual patients.

In Foel's article, progress in the development of CFTR modulator therapy was summarized very clearly. In this review, they focused on intestinal organoids as in vitro model for CF, understanding endogenous CFTR function, enabling CF therapy, drug developments and treatment optimization in a personalized manner [43]. It is also very pleasing and surprising to learn that CRISPR/Cas9 genomic editing system may be able to correct the mutation of CFTR in the intestinal organoids. This provides novel ideas on gene correction concepts for hereditary defects [44].

Male reproductive system organs have been developed as organoids, including the prostate and testis too. Undescended testis (Cryptorchidism) is one of the most common anomalies in pediatric surgery that is about 2–3% of babies have undescended testicles at birth which more common in premature babies. During the intrauterine life, gonads migrate from retroperitoneal area to the scrotums. If they fail to descend, fertility and neoplasms are related problems that may be seen in later stages of life. The testis is really an important organ responsible for sperm production and androgen synthesis. Researches on testicular organoids with spermatogenesis in vivo provide an excellent platform for understanding the mechanisms of mammalian spermatogenesis, and it might offer valuable personalized therapy for treating male infertility. Testis organoids enable studying samples in humans and other large animals where in vivo experiments are not possible, may provide a platform to design more precise therapeutic interventions, and allow modeling of testicular disease and malignancies as well [45, 46].

8.5 Fetal Organoids

Before discussing the potential applications of fetal organoids, it would be helpful to refer to some earlier studies we conducted at the Istanbul School of Medicine, Department of Pediatric Surgery. Although it is not about creating an organoid model, this intrauterine work is worth mentioning. We have shown in our previous studies that an avascular fetal intestinal segment may survive and continue to grow with an intact lumen in the subcutaneous space of a syngeneic host [47]. Our following study was designed to determine the absorptive capacity of the neo gut in vivo, without anastomosing the transplant to the host's intestine, and to evaluate its use as an accessory enteral segment. Intestinal segments taken from Wistar albino fetuses were transplanted subcutaneously into the abdominal wall of 20 Sprague-Dawley rats. The neo gut was started to grow and converted into a Thiry-Vella loop 2 weeks after transplantation. A test solution composed of 20% glucose and trophamine was perfused via the stomas; glucose and amino acid absorption gradients were calculated. Our data support the idea of using a fetal intestinal segment as an accessory feeding system alternative to intestinal transplantation [48]. Fetal tissue transplantation has gathered considerable interest among researchers dealing with organ transplantation.

It is possible that our early studies could provide inspiration for the fetal organoids that are currently very popular among scientists. In a more recent study, Chusilp postulated that intestinal organoids derived from fetal intestine and pluripotent stem cells are closely resemble immature intestinal epithelium. This fetal intestinal organoid can be used as intestinal tissue for regenerative medicine to damaged intestine or colon which provides a new hope of autologous tissue transplantation for infants in the near future [30]. Although the fetus is the youngest patient of pediatric surgeons, they are aware of the differences between developmental stages in the fetus and the postnatal physiology and anatomy of the infants. Therefore, organoid studies in children should be conducted with this in mind.

Organoids grown from amniotic fluid samples offer hope for studying congenital conditions and provide insight into how fetuses develop as well. Besides isolation of tissue-specific fetal stem cells and derivation of primary organoids obtained from termination of pregnancies, amniotic fluid may also be a source of cells from multiple developing organs. In a recent study, lung organoids were derived from amniotic and tracheal fluid cells of congenital diaphragmatic hernia fetuses [49]. Thanks to these advanced detailed studies, disease modeling and prenatal interventions are now possible allowing investigation of therapeutic tools and regenerative medicine strategies personalized to the fetus at clinically relevant.

In congenital diaphragmatic hernia cases (CDH), to get good results in treatment besides trying to correct the lung disorder, we must seek remedies to prevent compression in intrauterine life. So, the defect at the diaphragm must be repaired before the hypoplasia happen [23]. We must seek an answer to this question: Can lung be regenerated for the fetus and placed in utero?

CDH is a polygenic disorder associated with fetal lung compression and pulmonary hypoplasia at birth. Kunisaki and his colleagues induced pluripotent stem cell-derived lung organoids in an ex vivo model of the CDH fetal lung. They established a reproducible ex vivo model of lung development using transgene-free human-induced pluripotent stem cells generated from fetuses and infants with Bochdalek CDH [24]. Collectively, these data suggest both primary cell-intrinsic and secondary mechanical causes are the reasons of CDH lung hypoplasia and support the use of this stem cell-based approach for disease modeling in CDH.

8.6 Cancer Research in Pediatrics

Pediatric cancers are the leading cause of non-accidental death among children and represent children health priority. Organoids provide a reliable method for preclinical disease modeling, drug screening, gene editing and also provide the bases for gene and stem cell therapies for cancer. Researchers are now developing tumor organoids from pediatric cancers, such as neuroblastoma, Wilms tumor, teratomas, and rhabdomyosarcoma to study tumor biology and test therapies. Organoids are a new approach for improving pediatric cancer research. Patient-derived organoids in cancer medicine may have a great role in drug testing, biomarker discovery, and their potential for guiding personalized treatment [50].

In adults, relatively simple method to make research on bladder cancer comes in the form of established human bladder cancer-derived cell lines. A recent approach to cancer therapy has been the development of organoid systems that can apply to a number of different cancer types including prostate, colorectal, gastric, pancreatic, liver, and breast besides bladder [15].

Clinical interventions to preserve fertility and restore hormone levels in female patients with therapy-induced ovarian failure are insufficient. On the other hand, in Laronda's research, sexual maturity in mice following decellularized ovary transplant was demonstrated [51]. The recellularized grafts of ovariectomized mice provided data that could be used to drive future human transplants and have broader implications on the bioengineering of over carcinoma with endocrine function. Organoids also may have a role for precision medicine in testicular cancer. In spite of notable progress in establishing organoid models for adult tumors, there is a significant gap in equivalent research for pediatric tumors.

The adult tumors are often epithelial origin (such as in breast, lung, colon, and prostate cancers), but pediatric tumors on the other hand have an embryonal origin such as neuroblastoma, Wilms, teratoma, medulloblastoma, and retinoblastoma. Most of the embryonal tumors are highly malignant and consequently leading to poor prognoses. Despite the understanding of the underlying molecular landscape of these tumors that has helped in clinical advances to improve life expectancies of the children with neoplasms, there is still an emerging need to minimize morbidity and manage the long-term therapy effects by improving the studies on patient-derived organoids [52].

In a study about central nervous system tumors, it is attempted to reconstitute a brain tumor from a patient's primary tumor tissue in 3D in vitro models to match the parental tumor's molecular characteristics as much as possible. To date, the use of organoid models for pediatric brain tumors has been limited so far; however, there are only few publications on different tumor entities such as medulloblastoma, which may help with progress. In advancing medulloblastoma research and clinical management, uncertainty and scalability issues in the tumor microenvironment are critical to maximize their benefits [52].

Organoid-based drug research provides valuable insights into tumor evolution and drug resistance mechanisms and facilitates the rapid evaluation of drug efficacy and toxicity. By screening a range of chemotherapeutic agents, experimental drugs against tumor organoids and targeted therapies researchers can identify promising treatment protocols and facilitate their translation into clinical trials. This approach could enable a more rapid and ethically responsible transition from preclinical research to clinical practice in pediatric cancer treatment, thus improving the efficacy and safety of therapies offered to pediatric cancer patients.

8.7 The Future of Organoid Studies and Challenges

It is obvious that the regenerative medicine and organoids are useful to combat against very severe conditions and it is a great benefit to humankind. It has a great potential for drug assessments for personalized therapy, modeling congenital disorders for intrauterine developmental process, understanding etiopathology of the anomalies, surgical planning, regenerative medicine for alternatives to transplantation and cancer research. Studies are continuing with great speed and with effective technical advances. It looks very promising that in the near future, advancements in organoid technology will help to improve long-term outcomes for children. However, researchers must continue to anticipate the advantages and limitations of human organoids in patient-specific therapy, especially in children. There are still some technical and ethical limitations that further improvements are needed.

8.8 Bioethics

Bioethics is a branch of ethics that studies on philosophical, social, and legal issues arising in medicine and the life sciences. Before scientists learned how to engineer a patient's own cells, controversial fact was very disturbing that human embryos are used for research purposes. Researches on embryonic stem cell calls into question two moral principles that it was a big dilemma of weather the doctor's main mission to prevent or alleviate suffering from illness or uneasiness of embryonic stem cell researches. Naturally, both sides of the debate were interested in protecting human life, but the views are so different that risky consequences may occur. Nevertheless, in the following years, scientists learned how to engineer a patient's own cells so the discussions now lost its former tension. In spite of researchers' huge excitement for the potential of stem cells for curing diseases and conditions, there may be still some controversial facts [53]. It shows that further ethical research is needed especially on organoid transplantation and intrauterine procedures to help ensure the responsible development and clinical implementation of this technology in this field. It is not easy to approve and to accept immediately to use embryo cells for the other children or adults because of the respect to the value of human life; however, baby's own embryo cells can be used for itself to repair the congenital anomalies or acquired problems without much ethical objections.

Developments in science do not stand still. New horizons are constantly developing. Organs may not be grown not only from human cells, but also more different living organisms that even meat or vegetables, etc., may be produced by culturing various cells. It has a great potential not only for medicine but also for nutrition of the humankind that should be welcomed at least in less developed communities. Since it seems beyond our imagination, it creates fear and raises the issue of social negativities which may be against the scientists' improvements. Is it good or evil and is it right or wrong? How should they be regulated? Questions like these cannot be answered by science or ethical philosophy alone. As for me we must let the technology go that we can think the possible controversial facts in the future; otherwise, science remains as it is and no improvements for the human kind can be achieved. Rethinking on ethics and regulations in cell therapy and organoids seem necessary for a long time longer.

References

1. Laween M, et al. Engineering transplantable jejunal mucosal grafts using patient-drived orgonoids from children with intestinal failure. Nat Med. 2020;29:1593–601.
2. Langer R, Vacanti JP. Tissue engineering. Science. 1993;260:920–6.
3. Foglia RP, et al. Can fetal and newborn allografts survive in an immunocompetent host? J Pediatr Surg. 1986;21:608–12.
4. Somuncu ÖS. Decellularization concept in regenerative medicine. Adv Exp Med Biol. 2020;1212:71–85.
5. Rana D, et al. Development of decellularized scaffolds for stem cell-driven tissue engineering. J Tissue Eng Regen Med. 2017;11:942–65.
6. Sara M. Understanding short bowel syndrome: current status and future perspectives. Dig Liver Dis. 2020;52:253–61.
7. Kitano K, et al. Bioengineering of functional human induced pluripotent stem cell-derived intestinal grafts. Nat Commun. 2017;8:765.
8. Lee E, et al. Decellularized material as scaffolds for tissue engineering studies in long gap esophageal atresia. Expert Opin Biol Ther. 2017;17:573–84.
9. Urbani L, et al. Multi-stage bioengineering of a layered oesophagus with in vitro expanded muscle and epithelial adult progenitors. Nat Commun. 2018;9:4286.
10. Urbani L, et al. Long-term cryopreservation of decellularised oesophagi for tissue engineering clinical application. PLoS One. 2017;12:e179341.
11. Marrocco G, at al. Hypospadias surgery: a 10-year review. Pediatr Surg Int. 2004;20:200–3.

12. Somuncu ÖS, et al. Characterization and differentiation of stem cells isolated from human Newborn foreskin tissue. Appl Biochem Biotechnol. 2015;177:1040–54.
13. Huang JW, et al. Reconstruction of penile urethra with the 3-dimensional porous bladder acellular matrix in a rabbit model. Urology. 2014;84:1499–505.
14. Versteegden LRM, et al. Tissue engineering of the urethra: a systematic review and meta-analysis of preclinical and clinical studies. Eur Urol. 2017;72:594–606.
15. Banerjee S, Southgate J. Bladder organoids: a step towards personalised cancer therapy? Transl Androl Urol. 2019;8:300–2.
16. Tekguc M, et al. Kidney organoids: a pioneering model for kidney diseases. Transl Res. 2022;250:1–17.
17. Bas-Cristóbal Menéndez A, et al. Creating a kidney organoid-vasculature interaction model using a novel organ-on-chip system. Sci Rep. 2022;12:20699.
18. Hung SH, et al. Preliminary experiences in trachea scaffold tissue engineering with segmental organ decellularization. Laryngoscope. 2016;126:2520–7.
19. Elliott MJ, et al. Stem-cell-based, tissue engineered tracheal replacement in a child: a 2-year follow-up study. Lancet. 2012;380:994–1000.
20. Demchenko A, et al. Lung organoids: current strategies for generation and transplantation. Cell Tissue Res. 2022;390:317–33.
21. Stabler CT, et al. Revascularization of decellularized lung scaffolds: principles and progress. Am J Phys Lung Cell Mol Phys. 2015;309:1273–85.
22. Zhou H, et al. Bioengineering human lung grafts on porcine matrix. Ann Surg. 2017;267:590–8.
23. Okay ST, et al. The management Pediatric surgery of Bochdalek congenital diaphragmatic hernia: a single center experience and review of the literature. Eur Res J. 2020;6:138–44.
24. Kunisaki SM, et al. Human induced pluripotent stem cell-derived lung organoids in an ex vivo model of the congenital diaphragmatic hernia fetal lung. Stem Cells Transl Med. 2021;10:98–114.
25. Mazza G, et al. Decellularized human liver as a natural 3D-scaffold for liver bioengineering and transplantation. Sci Rep. 2015;5:13079.
26. Wai AWY, et al. Human liver organoids to predict the outcome of Kasai portoenterostomy. J Pediatr Surg. 2024;2025:161686.
27. Somuncu ÖS, et al. In vitro artificial skin engineering by decellularized placental scaffold for secondary skin problems of meningomyelocele. J Clin Neurosci. 2018;59:291–7.
28. Duisit J, et al. Decellularization of the porcine ear generates a biocompatible, nonimmunogenic extracellular matrix platform for face subunit bioengineering. Ann Surg. 2018;267:1191–201.
29. Hopkins RA, et al. Bioengineered human and allogeneic pulmonary valve conduits chronically implanted orthotopically in baboons: hemodynamic performance and immunologic consequences. J Thorac Cardiovasc Surg. 2013;145:1098–107.
30. Chusilp S, et al. Intestinal organoids in infants and children. Pediatr Surg Int. 2020;36:1–10.
31. Brendan C, et al. Paediatric gastric organoids as a tool for disease modelling and clinical translation. Pediatr Surg Int. 2021;37:317–24.
32. Clevers H. Modeling development and disease with organoids. Cell. 2016;165:1586–97.
33. Lotakis DM, et al. A pilot study: transcriptional profiling, functional analysis, and organoid modeling of intestinal mucosa in Hirschsprung disease. J Pediatr Surg. 2023;58:1164–9.
34. Hu X, et al. Necrotizing enterocolitis: current understanding of the prevention and management. Pediatr Surg Int. 2024;40:32.
35. Salman, et al. The effect of surgical trauma on the bacterial translocation from the gut. J Pediatr Surg. 1992;27:802–4.
36. Senger S, et al. Human fetal-derived enterospheres provide insights on intestinal development and a novel model to study necrotizing enterocolitis (NEC). Cell Mol Gastroenterol Hepatol. 2018;5:549–68.
37. McCracken KW, et al. Generating human intestinal tissue from pluripotent stem cells in vitro. Nat Protoc. 2011;6:1920–8.

38. McCracken KW, et al. Modelling human development and disease in pluripotent stem-cell-derived gastric organoids. Nature. 2014;516:400–4.
39. Jones BC, et al. Paediatric gastric organoids as a tool for disease modelling and clinical translation. Pediatr Surg Int. 2021;37:317–24.
40. Idowu S, et al. Gastric organoids: advancing the study of H. pylori pathogenesis and inflammation. Helicobacter. 2022;27:228–39.
41. Barkauskas CE, et al. Lung organoids: current uses and future promise. Development. 2017;144:986–97.
42. Dekkers JF, et al. A functional CFTR assay using primary cystic fibrosis intestinal organoids. Nat Med. 2013;19:939–45.
43. Foel D, at al. Intestinal organoids for cystic fibrosis research. J Cyst Fibros. 2020;19:60–4.
44. Schwank G, et al. Functional repair of CFTR by RISPR/Cas9 in intestinal stem cell organoids of cystic fibrosis patients. Cell Stem Cell. 2013;13:653–8.
45. Nengzhuang W, et al. A brief history of testicular organoids: from theory to the wards. J Assist Reprod Genet. 2022;39:1423–31.
46. Zhang D, et al. Establishment and characterization of testis organoids with proliferation and differentiation of spermatogonial stem cells into spermatocytes and spermatids. Cells. 2024;13:1642.
47. Salman FT, et al. Fetal intestinal transplantation in rats. Turk J Med Biol Res. 1990;1:167–9.
48. Güvenç BH, Salman FT, et al. Fetal intestinal transplant as an accessory enteral segment. Pediatr Surg Int. 1997;12:367–9.
49. Gerli MFM, et al. Single-cell guided prenatal derivation of primary fetal epithelial organoids from human amniotic and tracheal fluids. Nat Med. 2024;30:875–87.
50. Tong L, et al. Patient-derived organoids in precision cancer medicine. Med. 2024;5:1351–77.
51. Laronda M, et al. Initiation of puberty in mice following decellularized ovary transplant. Biomaterials. 2015;50:20–9.
52. Lampis S, et al. Organoids as a new approach for improving pediatric cancer research. Front Oncol. 2024;21:1414311.
53. de Jongh D, Massey EK, et al. Organoids: a systematic review of ethical issues. Stem Cell Res Ther. 2022;13:337.

Chapter 9
Organoid Technology in Cardiology for Disease Modeling and Treatment

Yahya Dağcan Bıçakçı and Irem Turkmen

Abstract Cardiovascular disease (CVD) is the leading cause of death globally, highlighting the urgent need for more accurate and human-relevant research models. Traditional in vitro and animal models have proven to be insufficient in replicating the structural and functional complexity of the human heart. Recent advances in stem cell biology and tissue engineering have facilitated the development of human cardiac organoids—three-dimensional, multicellular constructs derived from pluripotent stem cells that can recapitulate key features of cardiac development and disease. This chapter reviews the foundations, innovations, and limitations of cardiac organoid technology, exploring the mechanisms of organoid formation through self-organization, bioprinting, and tissue engineering. It also outlines different types of cardiac organoids and discusses their emerging applications in modeling inherited and acquired heart diseases, drug toxicity screening, and personalized medicine. While challenges remain, particularly regarding vascularization, maturity, and scalability, continuous improvements are positioning cardiac organoids as transformative tools in cardiovascular research. The chapter aims to provide a comprehensive overview of this cutting-edge field and highlight future directions for clinical and translational applications.

Keywords Cardiovascular disease · Cardiac organoids · Heart diseases · Drug toxicity screening

Y. D. Bıçakçı (✉)
Department of Cardiology, University of Health Sciences, Trabzon Medical Faculty, Trabzon, Turkey

I. Turkmen
Department of Cardiology, Istanbul University of Health Sciences Mehmet Akif Ersoy Thoracic and Cardiovascular Surgery Training and Research Hospital, Istanbul, Turkey

O. S. Somuncu (ed.), *Organoids for Preclinical Disease Modeling and Personalized Medicine*, Stem Cell Biology and Regenerative Medicine 78,
https://doi.org/10.1007/978-3-032-10024-5_9

9.1 Introduction

Cardiovascular disease (CVD) remains to be the leading cause of mortality worldwide, accounting for approximately 20.5 million deaths globally in 2021, representing nearly one-third of all global deaths [1]. Despite advances in the prevention and therapy, the burden of CVD continues to rise due to aging populations [2], sedentary lifestyles [3], diabetes [4], and obesity [5]. Traditional approaches in cardiovascular research have largely relied on animal models and 2D cell cultures, both of which suffer significant limitations. Rodent models, for instance, poorly replicate human cardiac physiology, particularly in areas such as ion channel regulation, drug metabolism, and electrophysiological conduction [6]. Moreover, standard monolayer cultures lack the three-dimensional structure and cellular complexity of native cardiac tissue [7].

Despite their anatomical and physiological similarities to humans, large-animal models face major translational barriers. Surgical protocols vary widely, experimental details are often under-reported, and rigorous study design, such as randomization, blinding, and sample size calculation, is frequently lacking. These issues, along with heterogeneity in species selection and infarct induction techniques, undermine reproducibility and reduce the predictive value of preclinical findings for human clinical outcomes [8]. Due to the poor translatability of traditional systems, there is growing interest in the development of in vitro, human-specific cardiac models that better recapitulate the microenvironment, structure, and function of the human heart. This interest aligns with the broader shift toward personalized medicine, which requires models that can replicate patient-specific phenotypes and predict individual drug responses. These demands have accelerated the emergence of next-generation platforms such as engineered heart tissues (EHTs) and cardiac organoids—technologies that aim to more faithfully model the complexity of human cardiac tissue.

9.2 Cardiac Organoids: Foundations and Innovations

The advent of cardiac organoids marks a transformative innovation in cardiovascular research, bridging the gap between simplistic in vitro systems and the complexity of the human heart [9]. Unlike engineered heart tissues (EHTs), which rely on external scaffolds and predefined molds, cardiac organoids emerge through spontaneous self-organization driven by morphogenetic signaling pathways such as Wnt, BMP, and FGF. This intrinsic capacity to mimic early cardiac development allows organoids to recapitulate key morphogenic events like heart tube formation and looping. These constructs often display spontaneous and synchronous contractions and include multiple cell types of cardiac origin, such as cardiomyocytes, endothelial cells, and fibroblasts, making them uniquely suited for studying early development and congenital heart disease [10].

Additional murine heart organoids created with FGF4 and ECM support, structures progress from cardiac crescent to heart tube and looping stages in a defined sequence, closely mirroring embryonic development [11]. This suggests that heart organoids are not just beating clusters of cells—they are spatially and temporally organized systems, capable of self-patterning in ways that mimic in vivo embryogenesis.

Progress in organoid maturation has further enhanced their physiological relevance. Through the application of biophysical stimuli—such as electrical pacing, mechanical stretch, and extracellular matrix conditioning—researchers have developed more mature cardiac tissues that exhibit aligned sarcomeres, calcium cycling, force generation, and even early T-tubule development [12]. Metabolic interventions using fatty acid-based media and long-term culture approaches have also contributed to advancing cardiomyocyte function toward a more adult-like phenotype [13]. These advancements suggest cardiac organoids are not only useful for developmental biology but also for drug discovery, disease modeling, and regenerative medicine applications.

Despite these remarkable strides, cardiac organoid technology still faces substantial limitations. A major challenge is their structural and functional immaturity—most organoids more closely resemble fetal rather than adult myocardium. Even with enhanced protocols, achieving the electrophysiological properties and contractile strength of mature human heart tissue remains difficult. Furthermore, the absence of a functional perfusable vascular network limits nutrient and oxygen diffusion to the core of larger organoids, often resulting in necrotic centers [14, 15]. Another significant limitation lies in the reproducibility of these systems. Because self-organization is inherently stochastic, batch-to-batch variability is common, complicating scalability and limiting the use of organoids in high-throughput drug screening [15, 16].

In terms of cellular complexity, current cardiac organoids do not yet fully replicate the multicellular interactions of native myocardium. While they include cardiomyocytes and endothelial cells, they often lack critical components such as the cardiac conduction system, autonomic innervation, and immune cells like macrophages, all of which are essential for accurately modeling inflammatory or electrophysiological diseases [17]. Additionally, organoids are generally cultured in static environments, lacking exposure to the mechanical forces and hemodynamic shear stress present in vivo [18]. This absence impairs proper myofibrillar organization and limits the physiological modeling of load-induced cardiac remodeling.

Nonetheless, researchers are actively addressing these limitations. Innovations such as microfluidic perfusion systems are being integrated to mimic blood flow and mechanical stimulation, and multi-lineage co-culture techniques are under development to incorporate immune and neural components [19, 20]. Coupled with advances in CRISPR-Cas9 gene editing and patient-derived iPSCs, these evolving models hold promise for disease-specific applications, such as modeling arrhythmogenic cardiomyopathy or evaluating personalized drug responses [21]. Moreover, image-based machine learning tools are increasingly used to monitor and quantify organoid development, helping to reduce variability and improve data robustness [22].

9.3 Types of Human Cardiac Organoids

The human heart exhibits significant structural and functional complexity, composed of various cell types such as cardiomyocytes and non-cardiomyocytes [12]. Cardiomyocytes, the primary contractile cells of the heart, possess limited regenerative capacity and are incapable of self-repair following injury. However, cardiomyocytes cultured in conventional two-dimensional systems fail to replicate the architectural and physiological characteristics of native myocardium. Non-cardiomyocyte populations, including endothelial cells, smooth muscle cells, fibroblasts, neurons, and immune cells, are embedded within a three-dimensional extracellular matrix (ECM) that provides structural support, signaling, and biomechanical alignment. Therefore, human cardiac organoids (hCOs) should strive not only to mimic the heart's functional output but also to reconstruct its cellular diversity and ECM-based architecture [12]. These organoids can be constructed using human embryonic stem cells, adult stem cells, or induced pluripotent stem cells, depending on the intended application (Table 9.1).

9.4 Methods of Cardiac Organoid Construction

Cardiac organoids are developed using principles of self-organization, directed differentiation, and tissue engineering. Three general strategies have emerged: tissue engineering methods, self-organization methods, and 3D bioprinting approaches (Fig. 9.1) [21]. One of the most common methods involves aggregating human pluripotent stem cells (hPSCs) in ultra-low adhesion microwells or suspension cultures, followed by stepwise modulation of Wnt signaling to induce mesodermal and cardiac specification. These aggregates self-organize into three-dimensional spheroids with visible contractility within 7–10 days [6, 12].

Embryoid body (EB)-based protocols offer an alternative method where hPSCs are first allowed to aggregate spontaneously, then exposed to cardiogenic factors to promote multilineage cardiac differentiation. Although these organoids show heterogeneous cell composition, they often display variability in size and structure, making them less suitable for uniform screening [6, 21]. A more engineered approach incorporates cardiac progenitors into hydrogels such as Matrigel or fibrin to provide biomechanical support and improve structural organization. Some models also apply electrical or mechanical stimulation to enhance maturation and functional performance [6, 21].

Table 9.1 Table summarizing key types of human cardiac organoids, based on recent peer-reviewed research, along with their defining features and key outcomes

Organoid type	Key features	Representative findings/ outcomes
Self-assembling heart organoids [23]	Wnt-based differentiation; epicardial layer near exterior	Multi-lineage organoids modeling heart development; useful for congenital disease modeling (e.g., diabetes-induced CHD)
Chambered cardiac organoids (CCOs) [24]	Form internal chambers; multiple lineage identity (CMs, endothelium, smooth muscle)	Upregulation of endothelial and smooth-muscle gene markers by day 14; transcriptomic alignment with developmental cardiac lineages
Patterned human primitive heart organoids (EMM/ chambered + proepicardial) [25]	Anterior-posterior heart field patterning; atrial/ventricular differentiation; proepicardial structures	Defined proportions of ventricular/atrial/ epicardial lineage clusters; recapitulate morphogenic patterning
Human Vascular Cardiac Organoids (hVCOs) [26]	Include vascular endothelial and fibroblast components	Enhanced CM maturation; integration of rudimentary vascular elements
Sympathetic neuron-innervated Assembloids (hSCAs) [26]	Incorporate neural crest/sympathetic neurons with chamber structure	Enables study of autonomic regulation of cardiac rhythm; arrhythmogenesis modeling
Elongating Multi-Lineage Organized Cardiac (EMLOC) gastruloids [26]	Highly complex: human chamber formation + vascular + neuronal integration	Model early cardiogenesis, chamber emergence, vascular neural interactions; recapitulate heart development at transcriptomic level

Recent developments include vascularized cardiac organoids produced by co-culturing cardiac cells with endothelial progenitors or embedding the organoids in microfluidic systems. These models better replicate in vivo oxygen and nutrient gradients and are suitable for studying ischemia and angiogenesis [6, 12]. Developmental cardiac organoids aim to mimic early heart formation and morphogenesis, providing insights into congenital heart defects and embryonic cardiac signaling pathways [12].

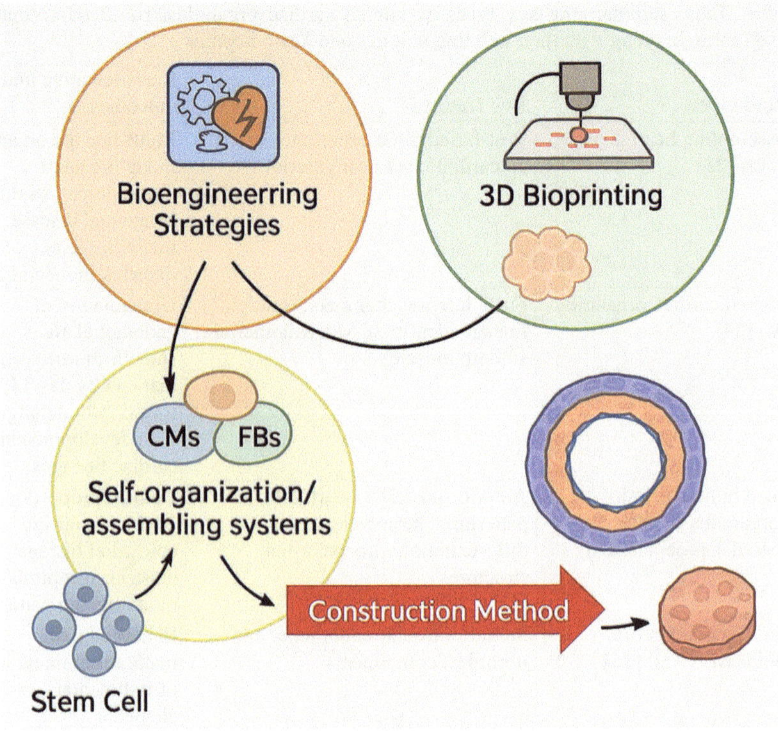

Fig. 9.1 Tissue engineering method, self-organization method, and 3D printing methods are the main three methods that are used for classification of hCOs construction [21]

9.5 Applications of Cardiac Organoids in Disease Modeling

Cardiac organoids serve as versatile platforms for modeling a spectrum of cardiovascular diseases. One prominent application is in genetic cardiomyopathies, where patient-specific iPSC-derived organoids carrying mutations can recapitulate disease hallmarks such as arrhythmia, fibrosis, and contractile dysfunction [6, 27]. These models enable mechanistic studies and preclinical drug testing within a human genetic context [21, 28]. Developmental cardiac organoids are also used to simulate embryonic heart development, including stages such as mesoderm induction and heart tube formation. These systems provide a platform to study congenital heart defects and investigate the effects of teratogenic agents [6, 12, 29].

Models incorporating vascular elements are increasingly employed to study ischemic heart disease and microvascular dysfunction. The inclusion of endothelial and smooth muscle cells improves diffusion dynamics and facilitates the investigation of ischemia-reperfusion injury and angiogenic therapies [6, 12, 30]. Additionally, organoids have shown utility in predicting cardiotoxic responses to pharmacological agents. By using genetically diverse iPSC lines, researchers can evaluate inter-individual differences in drug sensitivity, making cardiac organoids an important tool in safety pharmacology and personalized medicine [6, 21, 31].

9.6 Application of Cardiac Organoids for Downstream Molecular Assays

As previously mentioned, hCOs offer a robust and scalable platform for a wide range of molecular analyses aimed at dissecting cardiac development, disease mechanisms, and therapeutic responses. Because they contain multiple cardiac-relevant cell types, such as cardiomyocytes, endothelial cells, and fibroblasts, organized in a 3D context, these organoids closely recapitulate native heart tissue at both the structural and molecular levels. This makes them ideal for conducting transcriptomic, proteomic, epigenomic, and metabolomic assays.

9.6.1 Transcriptomics and Gene Expression Profiling

Bulk and single-cell RNA sequencing (scRNA-seq) are commonly applied to cardiac organoids to profile cellular heterogeneity, developmental stage-specific gene expression, and disease-associated transcriptional signatures. For example, scRNA-seq has been used to map lineage trajectories during mesoderm induction and early heart tube formation, as well as to reveal disruptions in gene expression caused by patient-specific mutations (e.g., *NKX2.5*, *MYH7*) [32, 33]. These techniques also enable comparison of healthy versus diseased organoids in response to stressors such as hypoxia, chemotherapeutic agents, or metabolic overload [34].

Using bulk RNA-sequencing across four distinct human cardiac organoid (hCO) fabrication methods, the authors demonstrated that variations in protocol result in significant differences in organoid maturity and cellular heterogeneity compared to 2D/3D hiPSC-derived cardiomyocytes and fetal/adult myocardium samples. Through PCA, k-means clustering, and enrichment analyses (GSEA, GSVA), they found that fabrication method directly shapes transcriptomic profiles—suggesting strategic protocol adjustments can standardize hCO maturity and functionality for targeted applications [35]. The lack of standardized protocols across research groups further contributes to inconsistency in outcomes.

9.6.2 Epigenetic and Chromatin Accessibility Studies

Studies utilizing assay for transposase-accessible chromatin with sequencing (ATAC-seq) have revealed dynamic changes in chromatin accessibility during differentiation of human pluripotent stem cells (hPSCs) into cardiac organoids, particularly at loci associated with cardiac development genes such as *NKX2–5*, *GATA4*, and *TBX5* [36]. By integrating ATAC-seq with scRNA-seq, researchers have reconstructed gene regulatory networks that guide early mesoderm patterning and chamber-specific differentiation.

In disease modeling, patient-derived cardiac organoids harboring mutations in key regulators (e.g., *LMNA*, *TBX20*) exhibit altered enhancer accessibility and transcription factor occupancy [37, 38]. For instance, a study applying CUT&RUN profiling to isogenic iPSC-derived cardiac organoids demonstrated that haploinsufficiency of *NKX2–5* led to disrupted chromatin architecture at cardiac enhancers and misexpression of downstream effectors involved in contractility and conduction [39].

These epigenomic tools also enable assessment of environmental perturbations. Organoids exposed to chemotherapeutics or metabolic stress show altered histone acetylation and methylation patterns, which can be linked to changes in cellular identity, stress response genes, and fibrotic remodeling [40]. Thus, chromatin profiling serves as a valuable readout for evaluating both developmental fidelity and pathological remodeling in organoid-based models.

Altogether, the integration of epigenomic profiling into cardiac organoid research is enabling mechanistic dissection of non-coding regulation, disease epigenetics, and drug-epigenome interactions, greatly expanding the functional utility of these platforms in basic science and translational medicine.

9.6.3 Metabolomic Studies

Using LC–MS metabolomics, a recent study profiled intracellular metabolites across early cardiac lineage stages—from pluripotent stem cells through mesoderm induction to contracting spheroids (~day 15). They found large-scale shifts in metabolic pathways, with significant upregulation of TCA cycle intermediates, NAD, and creatine and downregulation of glycolytic intermediates. This reflects a metabolic transition from proliferative, anabolic metabolism toward oxidative energy production consistent with early cardiomyogenesis in 3D hCO-like models [41].

Another study examined cardiac organoids that were cultured in varying palmitate and glucose concentrations. Addition of palmitate (10–100 μM) increased contractile force, reduced progenitor proliferation, and elevated maturation markers like MLC2v, demonstrating how metabolic substrates regulate organoid function [42]. These metabolomic profiles differentiate early vs. late-stage differentiation, providing a roadmap for designing maturation protocols and metabolic assays in cardiac organoid research.

9.7 Conclusion and Future Directions

Cardiac organoid technology represents a significant advancement in the field of cardiovascular research, offering a human-specific, three-dimensional platform that bridges the gap between conventional in vitro systems and the physiological complexity of native heart tissue. These models have demonstrated great potential in replicating developmental processes, modeling genetic and acquired cardiac diseases, and enabling personalized drug screening. The integration of patient-derived iPSCs, genome editing technologies, and microengineering techniques has enhanced the physiological relevance of these constructs and opened new avenues for translational applications.

Despite this progress, several challenges remain. Achieving structural and functional maturity that faithfully mimics adult myocardium is still an ongoing hurdle. Limitations in vascularization, reproducibility, and multicellular complexity constrain their utility in long-term studies and high-throughput applications. Standardized protocols and scalable manufacturing processes will be critical for their adoption in clinical and industrial settings.

Looking ahead, the incorporation of perfusable vascular networks, immune and neural components, and electromechanical stimulation systems holds promise for developing more physiologically complete cardiac organoids. Furthermore, advancements in organoid-on-chip technologies and AI-assisted imaging and analysis are expected to improve modeling accuracy and predictive power. With continued interdisciplinary collaboration, cardiac organoids are poised to become indispensable tools in disease modeling, drug development, and the pursuit of regenerative cardiovascular therapies.

References

1. World Health Organization. Cardiovascular diseases (CVDs). Geneva: World Health Organization; 2021.
2. Chong B, et al. Global burden of cardiovascular diseases: projections from 2025 to 2050. Eur J Prev Cardiol. 2024;32:1001–15.
3. Onagbiye S, et al. Association of sedentary time with risk of cardiovascular diseases and cardiovascular mortality: a systematic review and meta-analysis of prospective cohort studies. Prev Med. 2024;179:107812.
4. Siam NH, et al. Diabetes mellitus and cardiovascular disease: exploring epidemiology, pathophysiology, and treatment strategies. Rev Cardiovasc Med. 2024;25:436.
5. Powell-Wiley TM, et al. Obesity and cardiovascular disease: a scientific statement from the American Heart Association. Circulation. 2021;143:e984–e1010.
6. Zhu Y, et al. Cardiac organoids: a 3D technology for disease modeling and drug screening. Curr Med Chem. 2024;31:4987–5003.
7. Mathur A, et al. In vitro cardiac tissue models: current status and future prospects. Adv Drug Deliv Rev. 2016;96:203–13.
8. Shin HS, Shin HH, Shudo Y. Current status and limitations of myocardial infarction large animal models in cardiovascular translational research. Front Bioeng Biotechnol. 2021;9:673683.

9. Lewis-Israeli YR, Wasserman AH, Aguirre A. Heart organoids and engineered heart tissues: novel tools for modeling human cardiac biology and disease. Biomolecules. 2021;11:1277.
10. Cho J, et al. From engineered heart tissue to cardiac organoid. Theranostics. 2022;12:2758–72.
11. Lee J, et al. In vitro generation of functional murine heart organoids via FGF4 and extracellular matrix. Nat Commun. 2020;11:4283.
12. Yang Z, et al. Cardiac organoid: multiple construction approaches and potential applications. J Mater Chem B. 2023;11:7567–81.
13. Horikoshi Y, et al. Fatty acid-treated induced pluripotent stem cell-derived human cardiomyocytes exhibit adult cardiomyocyte-like energy metabolism phenotypes. Cells. 2019;8:1095.
14. Zhang S, Wan Z, Kamm RD. Vascularized organoids on a chip: strategies for engineering organoids with functional vasculature. Lab Chip. 2021;21:473–88.
15. Peletier M, et al. Multicellular 3D models to study myocardial ischemia-reperfusion injury. Front Cell Dev Biol. 2024;12:1494911.
16. Lee SW, et al. Proposal for considerations during human iPSC-derived cardiac organoid generation for cardiotoxicity drug testing. Biomed Pharmacother. 2024;174:116511.
17. Zhao Y, et al. A platform for generation of chamber-specific cardiac tissues and disease modeling. Cell. 2019;176:913–927 e18.
18. Blazeski A, Garcia-Cardena G, Kamm RD. Advancing cardiac organoid engineering through application of biophysical forces. IEEE Rev Biomed Eng. 2025;18:211–30. https://doi.org/10.1109/RBME.2024.3514378.
19. Joddar B, et al. Engineering approaches for cardiac organoid formation and their characterization. Transl Res. 2022;250:46–67.
20. Thomas D, et al. Cellular and engineered organoids for cardiovascular models. Circ Res. 2022;130:1780–802.
21. Wang YH, et al. Advances in cardiac organoids. Front Biosci (Landmark Ed). 2023;28:221.
22. Spiller ER, et al. Imaging-based machine learning analysis of patient-derived tumor organoid drug response. Front Oncol. 2021;11:771173.
23. Lewis-Israeli YR, et al. Self-assembling human heart organoids for the modeling of cardiac development and congenital heart disease. Nat Commun. 2021;12:5142.
24. Ho BX, et al. Robust generation of human-chambered cardiac organoids from pluripotent stem cells for improved modelling of cardiovascular diseases. Stem Cell Res Ther. 2022;13:529.
25. Volmert B, et al. A patterned human primitive heart organoid model generated by pluripotent stem cell self-organization. Nat Commun. 2023;14:8245.
26. Paredes-Espinosa MB, Paluh JL. Synthetic embryology of the human heart. Front Cell Dev Biol. 2024;12:1478549.
27. Bezzerides VJ, Zhang D, Pu WT. Modeling inherited arrhythmia disorders using induced pluripotent stem cell-derived cardiomyocytes. Circ J. 2016;81:12–21.
28. Zhao D, Lei W, Hu S. Cardiac organoid - a promising perspective of preclinical model. Stem Cell Res Ther. 2021;12:272.
29. Kim H, et al. Progress in multicellular human cardiac organoids for clinical applications. Cell Stem Cell. 2022;29:503–14.
30. Voges HK, et al. Vascular cells improve functionality of human cardiac organoids. Cell Rep. 2023;42:112322.
31. Grimm FA, et al. A human population-based organotypic in vitro model for cardiotoxicity screening. ALTEX. 2018;35:441–52.
32. Galdos FX, et al. Combined lineage tracing and scRNA-seq reveals unexpected first heart field predominance of human iPSC differentiation. elife. 2023;12:e80075.
33. Friedman CE, et al. Single-cell transcriptomic analysis of cardiac differentiation from human PSCs reveals HOPX-dependent cardiomyocyte maturation. Cell Stem Cell. 2018;23:586–598 e8.
34. Zhang L, et al. Modelling myocardial ischemia/reperfusion injury with inflammatory response in human ventricular cardiac organoids. Cell Prolif. 2025;58:e13762.

35. Hyams NA, et al. Improving human cardiac organoid design using transcriptomics. Sci Rep. 2024;14:20147.
36. Liu Q, et al. Genome-wide temporal profiling of transcriptome and open chromatin of early cardiomyocyte differentiation derived from hiPSCs and hESCs. Circ Res. 2017;121:376–91.
37. Mozzetta C, Tedesco FS. Challenging the "chromatin hypothesis" of cardiac laminopathies with LMNA mutant iPS cells. J Cell Biol. 2019;218:2826–8.
38. Tang Y, et al. TBX20 improves contractility and mitochondrial function during direct human cardiac reprogramming. Circulation. 2022;146:1518–36.
39. Feng W, et al. Computational profiling of hiPSC-derived heart organoids reveals chamber defects associated with NKX2-5 deficiency. Commun Biol. 2022;5:399.
40. Papazoglou P, Peng L, Sachinidis A. Epigenetic mechanisms involved in the cardiovascular toxicity of anticancer drugs. Front Cardiovasc Med. 2021;8:658900.
41. Bobori SN, et al. Metabolic remodeling during early cardiac lineage specification of pluripotent stem cells. Meta. 2023;13:1086.
42. Mills RJ, et al. Functional screening in human cardiac organoids reveals a metabolic mechanism for cardiomyocyte cell cycle arrest. Proc Natl Acad Sci USA. 2017;114:E8372–81.

Chapter 10
Organoids for Modeling Skin and Hair Pigmentation

Basak Tahmaz

Abstract For decades developing and advancing an accurate structural model of human skin is a significant goal for researchers. Skin is the outermost barrier, what shields us, a displayer to the outside world. Pathophysiology and management of skin diseases, investigating aging processes and cosmetic interventions, are some of the main ideals of dermatology. Organoid technology has come to the scene as the latest aspiring tool in dermatological research. Here, we collected a panel of papers for understanding the purposes of skin organoid technology on hair and skin pigmentation and their other applications in regenerative and reconstructive medicine, cancer research, and therapeutical advancements. This chapter coalesced both skin and hair follicle organoids studies by mainly focusing on recent publications for possible applications. Overall, this chapter discussed the usage of skin and hair follicle organoids, mainly on pigmentation, additionally on disease modeling, wound healing, enhancement of treatment options while exploring the future aspects, possible areas to expand.

Keywords Skin organoids · Hair follicle organoids · Pigmentation · Skin related diseases

10.1 Introduction

Skin, the largest organ of the human body, serves as an initial protecting barrier against UV light, pathogens, harmful chemicals, and mechanical damage while preventing excessive fluid loss, regulating temperature, and mediating sensation. Being a multilayered organ, consisting of the epidermis and dermis, with appendages like hair follicles and glands, skin sustains a remarkable regeneration capacity during

B. Tahmaz (✉)
Faculty of Medicine, Gazi University, Ankara, Turkey
e-mail: basaktahmaz@baskent.edu.tr

O. S. Somuncu (ed.), *Organoids for Preclinical Disease Modeling and Personalized Medicine*, Stem Cell Biology and Regenerative Medicine 78,
https://doi.org/10.1007/978-3-032-10024-5_10

homeostasis and injury. Many studies aim to generate an accurate in vitro skin model to be used in understanding developmental phases, disease modeling, drug screening, and reconstructive medicine.

Organoids can be broadly defined as three dimensional cultures derived from adult, fetal, or pluripotent stem cells. Researchers have shown that successful hair-bearing skin organoids can be generated from human-induced pluripotent stem cells [1, 2]. These three-dimensional cell cultures resemble the layered structure of the human skin, maintaining structural and functional relations between differing cell types, avoiding the limitations of animal models, computational simulations, and human cell cultures of primary strains [3].

10.2 Skin Organoids

Research on skin diseases is mostly carried out with murine models that lack the complexity of human skin. Skin organoids (SKOs) could be an excellent model to precisely study development. Lee et al. generated hair-bearing skin organoids resembling sphere-like cysts from homogenous population of human pluripotent stem cells (hPSCs) [2, 4]. hPSCs were sequentially treated with TGF-β inhibitors, BMP-4, low concentration of bFGF, BMP inhibitors, and high concentration of bFGF which induced development of cranial neural crest cells that differentiated into mesenchymal and neuroglial cells contributing to organoids dermis layer [2]. Hair germs started appearing at approximately 70 days [4]. Their data showed that while most organoid hair follicles were pigmented, they lacked the medulla layer which is characteristic of adult hair. SKOs produced lanugo-like hair, also including neuronal links around follicles maintaining a structure that resembles the 18-week human fetus. However, the model had its own limitations, for example, its cystic structure which restrained favorable hair growth [4]. A different study emphasizes the usage of air liquid interface is beneficial in reducing cyst formation and variation while enhancing hair follicle production. Air-liquid interface consists of a permeable membrane surrounded by air and liquid mediums on each side. The presence of sweat glands, sebaceous glands, sensory neuronal network, blood vessels, arrector pili muscles, adipose tissue, dermal papilla, hair follicles, hair shaft, Merkel cells, and pigmentation were seen in SKOs under bright-field images [5]. Air-liquid interface is additionally used as a medium for cutaneous organoid cultures, especially common for keratinization studies [6].

10.2.1 Hair Follicle Organoids

Hair follicle maturation, development, and pigmentation is modulated by different pathways such as TGF-β, FGF, and WNT signaling pathways [7, 8]. During developmental stages, hair follicles germs are formed by the interactions between mesenchymal and epidermal layers. In vitro hair follicle organoids (HFOs) were produced by the coculturing of epithelial and mesenchymal cells as aggregates [1]. After transplantation to nude mice, these hair follicle organoids regenerated effectively [1, 2]. In another study, hair follicle organoids were produced from mouse back skin. Mouse hair follicle organoids have a shorter life cycle than human HFOs which can provide faster development modeling and melanogenesis investigation [9]. In contrast to human HFOs, rodent-derived hair follicle organoids offer a more thorough approach. In the study carried out by the production of mouse follicle organoids, BMP4-treated nestin positive cells in the hair follicle bulge region differentiate into melanocytes. Under BMP4 stimulation, mouse HFOs showed enhanced pigmentation. BMP4 has a regulatory role in melanocyte function. It is the main signaling molecule, modifying transcription factors like MITF and melanogenic enzymes. BMP4 regulates hair follicle growth, enriching the numbers, length, pigmentation, and the size of hair follicles [9]. The regulatory role of BMP4 suggests a potential for treating pigmentation disorders. Successful generation of hair follicle organoids would provide a research platform to prevent hair loss and graying [9].

10.2.2 Skin and Hair Pigmentation and Organoid Applications

Pigmentation of the skin is critical for epidermal protection, acting as a mediator in infiltrating UV radiation. Moreover, it can be seen as evolutionary evidence that is rendered by environmental and genetic conditions [10]. The major pigment contributing to skin color is melanin. Melanin is synthesized by melanocytes of the basal layer of epidermis, in the specialized organelles named melanosomes. Melanosomes are further transported to keratinocytes where they are stored [11]. Disruption related to the components is associated with natural processes like hair graying, also hyperpigmentation or hypopigmentation disorders like age spots, vitiligo, and melasma [12].

One of many applications of organoids includes modeling of skin and hair pigmentation. Skin and hair color play a role in body image and cosmesis. Healthy hair is perceived as an indicator of well-being and youth. Hair color can alter physical appearance and as a result one's body image. As we age gray hair starts appearing. Hair graying is a process in which stem cells of melanocytes and melanocytes themselves were harmed, resulting in the deprivation of melanosome transfer and production. Tu et al. investigated the role of hair follicle organoids in pigmentation pathways. siRNA knockdown of genes affecting hair graying was examined on HFOs. Melanosome synthesis is regulated by Bcl2 and MITF, and the knockdown of these genes resulted with increased amount of gray hair [13].

Various forms of radiation result in different responses in melanogenesis. Solar radiation varies through different wavelengths and frequencies. Different types of light interact with biologic tissues and generate stimuli accordingly. Standardization of organoid culture models can be applied to estimate melanogenesis. Heterogeneous light sources with UVA, UVB, and visible light induce pigmentation in cutaneous organoid cultures from a fragment of human retro auricular skin [6]. Evaluating different behaviors of organoids promoted by diverse light sources promises improvement in photoprotection regimes [6]. A potential way to assess sunscreen formulations is through adopting skin organoids in future research.

UV radiation is a known carcinogen for humans. Pigmentation supplied by melanin is protective against photo damage and prevents skin cancers and aging. Supp et al. examined whether melanocyte admittance changed the response of engineered skin substitutes to UV light in vivo. Three different engineered skin substitutes (ESS), ESS from light skin phenotype, ESS from dark skin phenotype, and ESS deprived of melanocytes were tested under UV irradiation. ESS pigmentation levels were assessed by targeting melanocytes with TRYP1 monoclonal antibodies. Pigmentation levels did not increase in melanocyte-deprived ESS, meanwhile other ESS showed amplified pigmentation, significantly ESS with light melanocytes. Melanocyte bearing skin substitutes supplied reduced UV damage, promising reduced cancer risk [14].

10.3 Clinical Applications of Skin and Hair Follicle Organoids

10.3.1 Burns, Wound Healing, and Reconstruction

Small wounds heal closely to the original skin tissue when the progenitor cells of the basal epidermis are intact, while larger wounds may result in impaired regeneration and the loss of the appendages. Defective healing of larger wounds resulting with excessive scarring may cause loss of functions of the skin as well as cosmetic, social, and psychological problems. Scar morphology, significantly color and size, can alter patients' quality of life, even more than itch and pain of scarring [15].

Autologous skin grafting is the current standard management for large full-thickness burns. Injuries affecting larger percentage of total body surface area show greater risk of severe scarring, delayed healing, prolonged hospital stay that also leads to complications like surgical site infection and death. Other than autografts, engineered skin substitutes (ESS) as cultured epithelial autografts (CEA), which are fundamentally multilayers of autologous keratinocytes, and bilayer skin substitutes consisting of collagen-based polymers with autologous keratinocytes in addition to fibroblasts have been clinically tested. Since melanocytes were absent in the cultures (except the insignificant amount of "passenger melanocytes"), pigmentation was inadequate [16, 17]. Supplementation of selective melanocyte culture facilitated fully pigmented ESS [14, 16]. These classical in vitro human skin cultures may be simple equivalents; however, these models have limited relevance considering human skin is much more complex, including mature appendages that are crucial for function. Because of the limitations of two-dimensional skin substitutes,

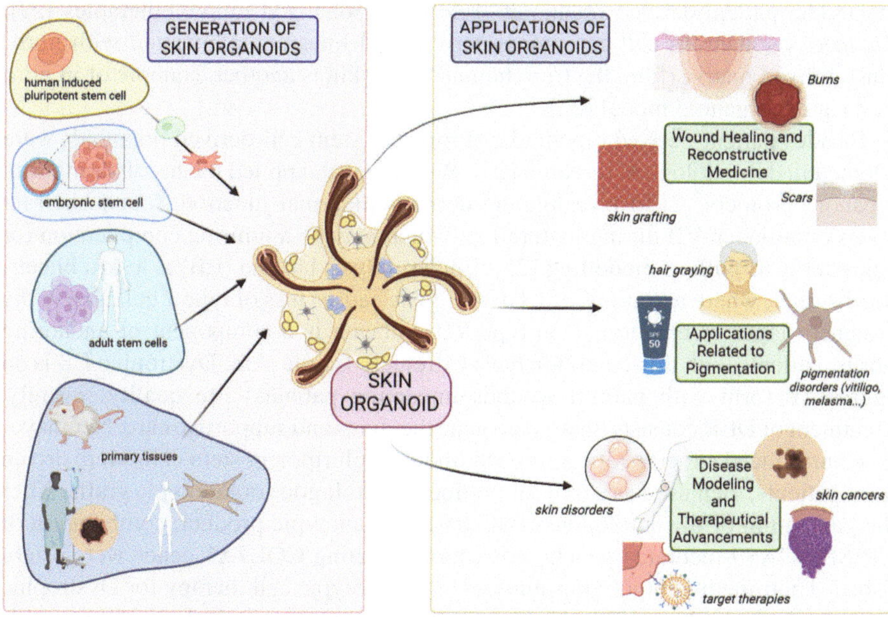

Fig. 10.1 Illustrations of skin organoid generation and applications. (**a**) Sources for skin organoid generation. (**b**) Applications of skin organoids. (Image created by BioRender)

three-dimensional skin organoids containing hair follicles could be an enhanced strategy for post-traumatic skin reconstruction. Restoring the pigmentation of skin and increased number of hair follicles within skin organoids may improve the cosmetic results and increase the rate of successful skin grafts (Fig. 10.1) [5].

10.3.2 Disease Modeling and Therapeutical Advancement

Skin organoids offer a promising alternative for modeling diseases meanwhile creating a unique approach to drug testing processes (Fig. 10.1). Patient-derived organoids can enlighten carcinogenesis as well as treatment options [18]. Skin cancer is one of the most prevalent cancers globally, with basal cell carcinoma (BCC), squamous cell carcinoma, and melanoma being the most common diagnosis [19–21]. Melanoma, malignant tumor of melanocytes, is a great example for patient derived organoids. Organoids derived from melanoma patients were produced by using different methods [22]. One method is planting melanoma patient-derived organoids (MPDOs) in collagen gel that were used to assess migration and chemotaxis of tumor infiltrating lymphocytes. The second method is to fix MPDOs in Matrigel which were used to determine the effectiveness of anti-PD-1 antibodies and other molecular therapeutic approaches. MPDOs showed similar characteristics to the parental tumor and highly immunosuppressive tumor microenvironment confirming melanoma patient-derived organoids' role in personalized immunotherapies [22]. Cutaneous squamous cell carcinoma-derived cell lines complemented with fibroblast-derived matrix (fdmDE) from human adult skin is another example of in vitro skin cancer organoid model [23].

Besides patient-derived organoids, pluripotent stem cell-derived organoids were adopted in dermatological research [24]. Ramovs et al. applied the hair-bearing skin organoid protocol [2] to investigating dermal-epidermal junction, found aberrant levels of collagen VII during maturation, which may be a limiting complication for epidermolysis bullosa modelling [25]. Epidermolysis bullosa (EB) is a skin blistering disorder, where mutations of COL7A1 gene causes loss of type 7 collagen in the basement layer of epidermis. As type VII collagen is a component of anchoring fibrils, patients with this condition have extremely fragile skin. Dystrophic EB is an aggressive form with painful wounds impacting patients' life quality severely. Treatment of DEB consists of targeted gene therapies and supportive care. Neumayer et al. introduced a therapeutic approach. Induced pluripotent stem cells were driven from patients, then used as a tool for producing autologous organotypic grafts. After the generation of the autologous iPSC-based organotypic product, combination of CRISPR/CAS9-mediated genetic correction targeting COL7A1 genes were established. Their method highlights autologous organotypic cell therapy for Dystrophic epidermolysis bullosa [26]. Usage of pluripotent stem cell-derived organoids extends to human pigmentation diseases. There are few suitable examples for in vitro modeling of pigmentation disorders of the skin. One of them is hyperpigmentation of Neurofibromatosis type 1. Neurofibromatosis type 1 is a monogenic neurocutaneous disorder expressing itself via a wide range of symptoms including dermatological manifestations. Skin defects included in diagnostical criteria include hyperpigmentation, café au lait spots, axillary and inguinal freckling, and neurofibromas. NF1 gene is responsible for neurofibromin production. Neurofibromin has a role in normal cell cycle progression. Decreased expression of NF1, and as a result neurofibromin, causes adjustments of cell cycle. Hyperpigmentation in Neurofibromatosis type 1 patients was modeled using melanocytes generated from their embryonic stem cells [27]. Their findings support that neurofibromin is a part of human melanogenesis as depletion of NF1 resulted increases in melanin synthesis. Moreover, role of ERK and cAMP signaling pathways was demonstrated along with pharmacological modifications of these pathways. In future research, hyperpigmentation modeling and management could be explored via three dimensional organotypic cultures derived from NF1 patients, enlightening pathophysiology of the disease, providing further options of enriching molecular treatment. In vitro skin organoid models mimic the progression of disease precisely, promising breakthroughs in skin disorder research, deciphering the underlying mechanisms and potential advancements in therapies (Fig. 10.1) [28].

10.4 Discussion

Organoids are reliable three-dimensional models for reproducing varying tissues in vitro. While different tissue-specific organoids were described in other studies, this chapter reviews the generation and diverging uses of skin organoids, especially its appendage, hair follicle, containing organoids. SKOs were developed in mediums such as Matrigel and air-liquid interface. Sources to produce SKOs can be summarized as human pluripotent stem cells, embryonic stem cells, adult stem cells, and other primary tissue cultures such as fibroblasts and keratinocytes, cell cultures collected from patients and animal tissues. Skin and hair follicle organoids may have some challenges during administration processes. Limitations of in vitro skin organoid models were reported as cystic structure leading to inadequate appendage formation, demanding equal microenvironment as in vivo, and if the organoid model originated from cells other than human tissue, they lacked close alignment of human skin.

Hair follicle germ containing aggregates can generate hair follicles when transplanted as grafts. In vitro hair follicle models are instrumental for studying hair follicle induction, assessing drug reactions on hair growth, and modeling of graying hair. Hair follicle organoids are the future of research related to hair.

Melanocytes are crucial for hair and skin pigmentation considering their essential function in synthesis and distribution of the pigment melanin. Melanin not only plays a role in pigmentation but also provides shielding from light-induced cellular damage. Understanding photocarcinogenesis is an important goal and using organoids to complete this goal is the future of carcinogenesis studies. Furthermore, enlightening pigmentation disorders can be carried out.

Skin organoids (SKOs) represent a rapidly advancing platform that recapitulates the complex architecture and cellular heterogeneity of human skin, enabling precise modeling of developmental and pigmentary processes, elucidation of disease pathways, and the refinement of regenerative and therapeutic strategies. Wounds like burns can lead to scars and other undesirable complications, especially if they are in larger size, exceeding skin's replicating ability. Advancing the recent standard treatment by applying organoids containing hair follicles and melanocytes in grafting techniques could be the future of reconstructive medicine.

Gaining insight into the processes that lead to skin disorders, their genomic backgrounds, behaviors, triggers, and risk factors is the main course to prevent and treat them accordingly. Patient-derived organoids can be used as an approach for this intention. Melanoma patient-derived organoids were used to test molecular treatments for melanoma. Likewise, other disorders that skin organoids can function to elucidate are epidermolysis bullosa and neurofibromatosis type 1, which were explained in this article.

10.5 Conclusion

Skin organoids have a transformative potential since precise in vitro models of human skin are necessary for advancing dermatological research. Accurate methods of human skin organoid generation have pioneered in exploring skin architecture, maintenance, repair, disease modeling and treatment options. Organoid systems promise continuation of evolution. As skin organoid models sustain improvement, they offer unique solutions from regenerative medicine to targeted therapies.

References

1. Tu S, Kageyama T, Seo J, Zhou Y, Fukuda J. Development of in vitro hair pigmentation model using hair follicle organoids. J Biosci Bioeng. 2025;139:141–6.
2. Lee J, van der Valk WH, Serdy SA, Deakin C, Kim J, Le AP, et al. Generation and characterization of hair-bearing skin organoids from human pluripotent stem cells. Nat Protoc. 2022;17:1266–305.
3. Zhou L, Zhang X, Paus R, Lu Z. The renaissance of human skin organ culture: a critical reappraisal. Differentiation. 2018;104:22–35.
4. Lee J, Rabbani CC, Gao H, Steinhart MR, Woodruff BM, Pflum ZE, et al. Hair-bearing human skin generated entirely from pluripotent stem cells. Nature. 2020;582:399–404.
5. Sun J, Ahmed I, Brown J, Khosrotehrani K, Shafiee A. The empowering influence of air-liquid interface culture on skin organoid hair follicle development. Burns. Trauma. 2025;13:tkae070.
6. Olivatti TOF, Alcantara GP, Lemos A, Silva MGD, Miot HA. Standardization of organoid culture for evaluation of melanogenesis induced by UVB, UVA and visible light. An Bras Dermatol. 2020;95:46–51.
7. Chen D, Jarrell A, Guo C, Lang R, Atit R. Dermal beta-catenin activity in response to epidermal Wnt ligands is required for fibroblast proliferation and hair follicle initiation. Development. 2012;139:1522–33.
8. Kageyama T, Shimizu A, Anakama R, Nakajima R, Suzuki K, Okubo Y, et al. Reprogramming of three-dimensional microenvironments for in vitro hair follicle induction. Sci Adv. 2022;8:eadd4603.
9. Xie R, Li M, Wang X, Liu Z. BMP4 regulates differentiation of nestin-positive stem cells into melanocytes. Cell Mol Life Sci. 2025;82:41.
10. Jablonski NG. The evolution of human skin pigmentation involved the interactions of genetic, environmental, and cultural variables. Pigment Cell Melanoma Res. 2021;34:707–29.
11. Hearing VJ. Biogenesis of pigment granules: a sensitive way to regulate melanocyte function. J Dermatol Sci. 2005;37:3–14.
12. Nicolaidou E, Katsambas AD. Pigmentation disorders: hyperpigmentation and hypopigmentation. Clin Dermatol. 2014;32:66–72.
13. Tobin DJ. Human hair pigmentation–biological aspects. Int J Cosmet Sci. 2008;30:233–57.
14. Supp DM, Hahn JM, Lloyd CM, Combs KA, Swope VB, Abdel-Malek Z, et al. Light or dark pigmentation of engineered skin substitutes containing melanocytes protects against ultraviolet light-induced DNA damage in vivo. J Burn Care Res. 2020;41:751–60.
15. Oh H, Boo S. Assessment of burn-specific health-related quality of life and patient scar status following burn. Burns. 2017;43:1479–85.
16. Boyce ST, Lloyd CM, Kleiner MC, Swope VB, Abdel-Malek Z, Supp DM. Restoration of cutaneous pigmentation by transplantation to mice of isogeneic human melanocytes in dermal-epidermal engineered skin substitutes. Pigment Cell Melanoma Res. 2017;30:531–40.

17. Swope VB, Supp AP, Boyce ST. Regulation of cutaneous pigmentation by titration of human melanocytes in cultured skin substitutes grafted to athymic mice. Wound Repair Regen. 2002;10:378–86.
18. Dutta D, Heo I, Clevers H. Disease Modeling in stem cell-derived 3D organoid systems. Trends Mol Med. 2017;23:393–410.
19. Aggarwal P, Knabel P, Fleischer AB Jr. United States burden of melanoma and non-melanoma skin cancer from 1990 to 2019. J Am Acad Dermatol. 2021;85:388–95.
20. Lomas A, Leonardi-Bee J, Bath-Hextall F. A systematic review of worldwide incidence of nonmelanoma skin cancer. Br J Dermatol. 2012;166:1069–80.
21. Sathe NC, Zito PM. Skin cancer. Treasure Island: StatPearls. Ineligible Companies. Disclosure: Patrick Zito declares no relevant financial relationships with ineligible companies; 2025.
22. Ou L, Liu S, Wang H, Guo Y, Guan L, Shen L, et al. Patient-derived melanoma organoid models facilitate the assessment of immunotherapies. EBioMedicine. 2023;92:104614.
23. Berning M, Pratzel-Wunder S, Bickenbach JR, Boukamp P. Three-dimensional in vitro skin and skin cancer models based on human fibroblast-derived matrix. Tissue Eng Part C Methods. 2015;21:958–70.
24. Oceguera-Yanez F, Avila-Robinson A, Woltjen K. Differentiation of pluripotent stem cells for modeling human skin development and potential applications. Front Cell Dev Biol. 2022;10:1030339.
25. Ramovs V, Janssen H, Fuentes I, Pitaval A, Rachidi W, de Sousa C, Lopes SM, et al. Characterization of the epidermal-dermal junction in hiPSC-derived skin organoids. Stem Cell Reports. 2022;17:1279–88.
26. Neumayer G, Torkelson JL, Li S, McCarthy K, Zhen HH, Vangipuram M, et al. A scalable and cGMP-compatible autologous organotypic cell therapy for dystrophic epidermolysis bullosa. Nat Commun. 2024;15:5834.
27. Allouche J, Bellon N, Saidani M, Stanchina-Chatrousse L, Masson Y, Patwardhan A, et al. In vitro modeling of hyperpigmentation associated to neurofibromatosis type 1 using melanocytes derived from human embryonic stem cells. Proc Natl Acad Sci USA. 2015;112:9034–9.
28. Zhou R, Luo X, Chen S, Jia K, Yu H, Huang Z, et al. Innovative approaches in skin therapy: the rise of organoid cultivation. Chem Eng J. 2025;512:162248.

Chapter 11
Next-Generation Sequencing Methods Applied to Organoids in Cancer Research

Martin Egger

Abstract The integration of next-generation sequencing (NGS) has elevated organoid research significantly and was instrumental in making patient-derived organoids (PDOs) a promising tool for biomarker discovery, drug response analysis in vivo, and personalized treatment strategies. On one hand, high-throughput capabilities have made NGS invaluable in comparing organoids with parental tumors they were derived from, assess genetic stability over multiple organoid passages, and characterize rare subpopulations in organoids. On the other, organoids have the capability to increase sequencing depth and sensitivity compared to necrosis-prone tumor biopsies. This review highlights the applications, advantages, and limitations of NGS methods in genomics, transcriptomics, and epigenomics when applied to organoid-based cancer studies.

Keywords Next-generation sequencing · Cancer organoids · Genomics of cancer organoids · Transcriptomics of cancer organoids · Epigenomics of cancer organoids

11.1 Introduction

Organoids are three-dimensional, multicellular structures derived from stem cells that replicate the architecture and functionality of native tissues [1]. In recent years, organoids have become valuable models in cancer research, offering a more accurate representation of tumor biology compared to traditional two-dimensional cell cultures [2, 3]. These models are especially useful in drug testing, disease modeling,

M. Egger (✉)
Department of Radiation Oncology, Dana-Farber Cancer Institute, Boston, MA, USA

Broad Institute of Harvard and MIT, Boston, MA, USA

Hospital Center Biel, Spitalzentrum Biel, Department of Urology, Biel, Switzerland
e-mail: megger@broadinstitute.org

O. S. Somuncu (ed.), *Organoids for Preclinical Disease Modeling and Personalized Medicine*, Stem Cell Biology and Regenerative Medicine 78,
https://doi.org/10.1007/978-3-032-10024-5_11

and understanding cancer at a molecular level. Organoids derived from patient tissues closely reflect the genetic, transcriptomic, and epigenetic features of the parental tumor, making them an interesting platform for personalized medicine [4]. The rise of next-generation sequencing (NGS) technologies has significantly advanced cancer research by allowing high-throughput, cost-effective analysis of cancer genomes [5]. NGS provides comprehensive insights into the DNA, RNA, and epigenetic changes in cancer cells, enabling researchers to better understand tumor biology, identify new biomarkers, and develop targeted therapies [6, 7]. When applied to organoids, NGS technologies enhance the ability to study genetic variations, mutations, and molecular mechanisms that drive cancer.

A search for "organoids" and "next-generation sequencing" on PubMed shows a clear increase in publications in a similar time period for both terms, underscoring the growing trend of combining these technologies in cancer research (Fig. 11.1). Organoids, which more accurately reflect a tumor's heterogenous genetic profile than traditional cell-lines, allow for the detection of key mutations, gene expression

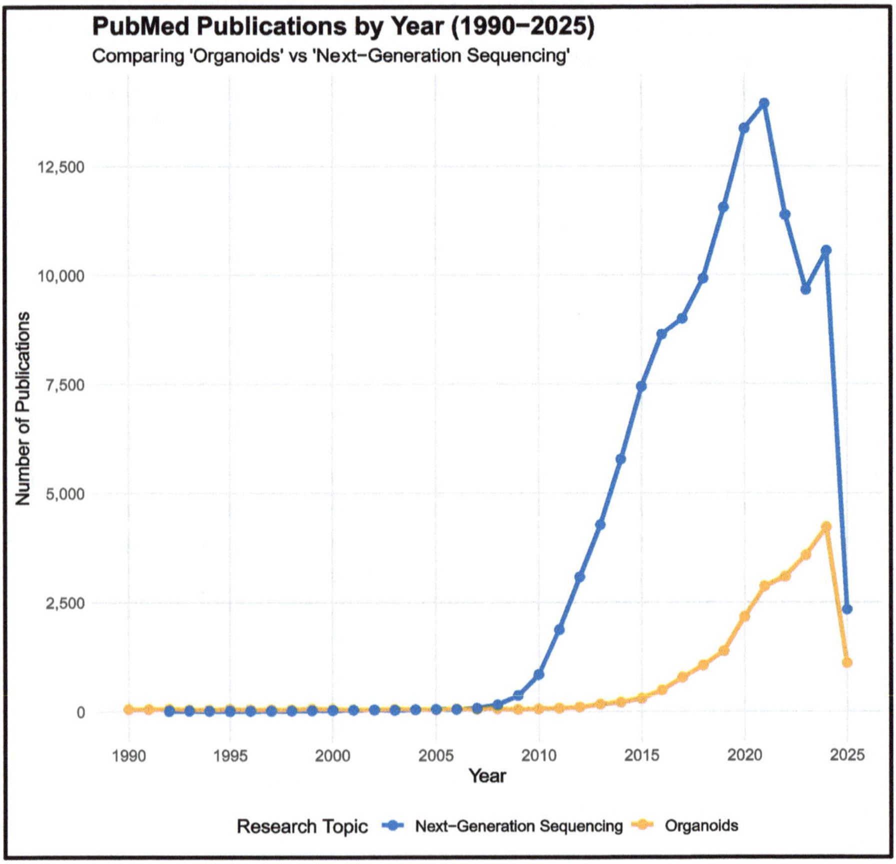

Fig. 11.1 PubMed search term "organoids" and "next-generation sequencing"

differences, and epigenetic alterations regulating gene expression, which are critical in tumorigenesis [8]. NGS technologies like whole-genome sequencing (WGS), whole-exome sequencing (WES), RNA sequencing (RNA-seq), and epigenomic techniques allow for a deeper understanding of cancer's molecular landscape and identify potential therapeutic targets. As this review will show, NGS has become invaluable for comparing organoids with their parental tumors, providing insights into their genetic similarities and differences. It also allows researchers to assess genetic stability across multiple organoid passages and identify rare subpopulations that contribute to tumor heterogeneity. Additionally, organoids enhance sequencing depth and sensitivity, offering a clearer genomic profile than necrosis-prone tumor biopsies. This review will focus on the integration of NGS in cancer research using organoid models, examining the advantages and limitations of these approaches. While other critical technologies like proteomics or metabolomics are beyond the scope of this review, it will instead explore genomic, transcriptomic, and epigenomic techniques, providing a detailed look at how NGS contributes to cancer research and personalized medicine.

11.2 Genomics

Various technologies have been used in organoid research to assess genomic variation and to compare the genomic characteristics to the tumor tissue of origin. Above all, the types of genome sequence variants generally fall into two main categories [9]. The first category accounts for only 1–2% of the genome and includes changes in exons or protein-coding regions of genes, such as loss-of-function (LoF) and gain-of-function (GoF) mutations. LoF variants are typically missense or nonsense variations (such as single-nucleotide polymorphisms or SNPs) that disrupt protein function or synthesis. While most variants are SNPs, they can also be insertions or deletions which can impair correct messenger RNA assembly and therefore protein function [10]. Non-coding regions of genes make up the majority of the genome with 98–99% of the human genome [11].

The function of these non-coding regions and their significance on the phenotype at the cellular level are apart from regulatory functions mostly unclear [12]. However, we do know that some variants and mutations in these non-coding regions of the genome can cause disease [13, 14]. This subsection will give an overview over the most commonly used technologies in organoid research and highlight their individual strengths and weaknesses.

11.2.1 Whole-Genome Sequencing

Whole-genome sequencing (WGS) captures the entire genome including the exome and the non-coding regions. WGS detects SNPs, structural variants (SVs), insertions and deletions (indels), copy number alterations (CNAs), and fusion genes and is widely recognized as the most comprehensive and unbiased method of genomic characterization within cancer studies [15, 16]. WGS offers the exploration of the full genomic landscape of cancer biopsy-derived organoids, allowing a complete picture of patterns of alterations and mutational signatures, which can highlight potentially targetable pathways [17]. The WGS platforms can be divided into short- and long-read protocols. Short-read sequencing is the most widely used method generating reads of <300 base pairs (bp) and is generally a fast, high-accuracy and sequencing depth at a relatively low cost [18]. Long-read sequencing on the other hand generates reads ranging from 10kbp to several mega-bases and is the preferred option to detect larger, more complex SVs and repeats [19].

Whole-genome sequencing (WGS) has been a pivotal method for genomic characterization in organoid research. It has been widely used to detect mutations, structural variants (SVs), insertions and deletions (indels), copy number alterations (CNAs), and fusion genes. WGS provides a comprehensive and unbiased view of the entire genome, including exonic and non-coding regions, which allows for an in-depth analysis of cancer genomics. WGS has been used to identify common driver mutations in various cancer types. For instance, in colorectal cancer, WGS was instrumental in demonstrating the accumulation of driver mutations such as WNT, TP53, ACVR2A, and BMP in long-term cultures of mismatch-repair-deficient PDOs, recapitulating carcinogenesis [20]. Similarly, WGS was used in lung cancer PDOs to identify missense mutations and oncogenic alterations in known cancer-associated genes, including LoF mutations, like TP53 and STK11, and GoF mutations such as KRAS and ERBB2 [21]. In addition to identifying mutations, WGS is also effective in comparing the genomic landscapes of PDOs with their parental tumors. For example, in neuroendocrine neoplasms (NENs), WGS demonstrated a high accuracy in classifying variants as somatic or germline in PDOs derived from multiple organs [22]. Another study on NEN-derived PDOs used WGS to show a high degree of similarity between the PDOs and their parental tumors, particularly in terms of tumor mutational burden (TMB), SNPs, and indels [23]. In pancreatic cancer, WGS was essential for detecting somatic mutations in a germline-corrected cohort using normal pancreatic tissue. The higher tumor purity and cellularity in PDOs compared to healthy tissue enabled more accurate detection of actionable genetic alterations, especially in pancreatic ductal adenocarcinoma, a carcinoma prone to high necrosis and low tumor purity in primary samples [24]. WGS also plays a critical role in assessing genetic stability over time. In liver cancer PDOs, chromosomal aberrations remained highly stable over multiple passages, supporting the potential of PDOs as a biobanking platform [25]. Further studies have shown that PDOs can serve as effective models for genetic analysis in other cancer types. For example, in breast cancer, WGS demonstrated that somatic mutations, CNVs, and DNA copy number alterations were retained in PDOs, and these organoids exhibited cleaner CNV profiles compared to parental tumor biopsies, making it easier to detect somatic mutations [26]. Additionally, to increase resolution to the single-cell level, single-cell dissociation followed by clonal organoid generation and WGS has been suggested as a method to achieve finer detail in genomic analyses [27]. While WGS is a widely used and valuable tool in organoid research, it comes with its limitations. Its higher cost compared to more targeted sequencing techniques, as well as the large amount of data it generates, can be challenging. The high computational burden may be mitigated through advanced bioinformatics solutions, such as virtual gene panels. These panels apply filters to WGS data, retaining only sequences from clinically relevant genes, which allows for faster computational processing and reduces the workload for data interpretation [28, 29].

11.2.2 Whole-Exome Sequencing

As opposed to WGS, whole-exome sequencing (WES) targets the protein coding regions only, the exome, comprising 1–2% of the whole genome. This limited approach is justified as multiple studies estimate that around 85% of disease-causing mutations, including cancer, are located in the exome [30]. This streamlined approach is a more targeted option to WGS at a much more cost-effective value due to its faster throughput, reduced computational burden, and high diagnostic yield [31]. Despite its substantial diagnostic potential at a relatively affordable cost, WES has several key limitations. Due to its capture of the protein-coding regions only, it can disregard clinically relevant intronic variants. In addition, WES has the limitation of inefficient capture of coding sequences, read depth, and alignment errors depending on the platform used for WES. Variant prioritization as a step of data analysis of WES will differ in laboratories, making the inclusion or exclusion of variants a biasing factor [32].

In pancreatic cancer, WES showed, similarly to WGS, that tumor purity is higher in PDOs compared to parental tissue, allowing for a higher sensitivity in calling clinically relevant somatic mutations. In one study, PDOs demonstrated the ability to detect a higher proportion (29%) of mutation calls compared to only 3% from the primary tumor samples [33]. This highlights that PDOs provide a clearer picture of the mutational landscape than the parental tissue in pancreatic cancer. WES can also be valuable for assessing mutational concordance between PDOs and their parental tumor tissues. For instance, in colorectal carcinoma, WES was used to identify tumor-specific somatic mutations, mutation rates per Mb, and CNVs and to match these with WES data from normal colorectal biopsies [34]. Additionally, in head and neck cancer, WES differentiated tumor PDOs from non-cancerous ones by classifying only PDOs carrying mutations in known tumor suppressor genes or oncogenes (e.g., TP53, NOTCH1, PIK3CA, FAT1, and APOB) as true tumor organoids [35]. WES also plays a role in monitoring genetic stability over time in organoid cultures. For example, in prostate cancer, WES was performed on both long-term and short-term organoid cultures to assess single-nucleotide variants (SNVs), small indels, and loss of heterozygosity (LOH). No genetic variants were detected in late organoid passages compared to early ones, suggesting genetic stability in PDOs over time [36]. Similarly, in liver cancer, WES was used to confirm that PDOs retained the parental tumor's mutational landscape over extended passages, with SNVs and indels remaining stable in both short- and long-term cultures [37]. In castration-resistant prostate cancer, WES was applied to catalog somatic and germline mutations, including SNVs and indels, identifying the most commonly mutated genes such as TP53, FOXA1, and PIK3R1 [38]. While WGS offers a more comprehensive picture of mutational signatures and the full genomic landscape, WES remains a scientifically viable and cost-effective alternative. As shown in multiple studies, WES provides useful insights, particularly for mutational analysis and genetic stability, making it widely used in organoid research projects.

11.2.3 Targeted Sequencing

Targeted sequencing represents a completely different approach of mutation scanning. While WGS and WES are important in defining a pathogenic variant that may be difficult to suspect or might not have been previously recorded, targeted sequencing focuses on mutations of genes known to be associated with disease. Targeted sequencing technologies selectively sequence desired DNA sequences of individual or multiple genes, known as gene panels. Focusing on gene panels and therefore on only a fraction of the genome significantly lowers the cost of sequencing and can even offer a higher depth of coverage of the genes of interest than WGS or WES, while also facilitating interpretation, which in some cases can result in a similar diagnostic yield [39–41].

In organoids, targeted sequencing can be used to differentiate tumor organoids from non-cancerous tissue organoids. For example, in hepatoblastoma, targeted sequencing of the CTNNB1 gene encoding for β-catenin, a known hepatoblastoma marker, was used to detect true tumor PDOs [42]. Similarly, in head and neck cancer organoids, targeted sequencing was a valid alternative to WES to distinguish cancerous from benign cells, sequencing known tumor suppressor and oncogenes (TP53, NOTCH1, PIK3CA, FAT1, and APOB) [35]. In another study on PDOs derived from gastrointestinal tumors, targeted sequencing was used first to detect the similarity of genomic alterations in PDOs and parental tissue with a defined gene panel of 151 cancer-related genes and second to detect targetable mutations [43]. The main limitation of targeted sequencing is its inability for pathogenic or targetable variant discovery. This however might not be the main aim of using sequencing in organoid research, with the exception of tumors with a high percentage of necrotic tissue, such as pancreatic ductal adenocarcinoma [24].

11.3 Transcriptomics

Assessing a tumor's transcriptome and therefore gene expression is essential in understanding its molecular biology and behavior such as drug response. Similar to DNA, RNA is divided into messenger RNA (mRNA) and non-coding RNA. As mRNA is translated into polypeptides and eventually proteins further downstream, sequencing technologies have focused on mRNA to assess gene expression. As opposed to genomics, gene expression gives us insight into the function of cells after maturation. Gene expression patterns are widely used in a variety of tumor entities as they directly impact the cellular phenotype and can be used to differentiate between disease entities [44] and to stratify cancers into molecular subtypes, which in some cases can predict outcome [45–47]. In performing RNA sequencing (RNA-seq) on PDOs and their parental tumor tissues, RNA-seq generally allows us to compare active and dormant cellular processes.

11.3.1 Bulk RNA Sequencing

Bulk RNA-seq is the most commonly used and widely available technique for exploring gene expression. One of its primary functions is to compare the gene expression profiles of PDOs with their parental tissues. In colorectal cancer, for example, bulk RNA-seq was used in a PDO biobanking project. Due to the PDO generation protocol, the organoids were enriched in epithelial cells, allowing for cleaner bulk RNA-seq results compared to tissue biopsies that could contain mesenchymal, endothelial, or immune cells. Organoids were derived from tumor biopsies and normal colon epithelium and then stratified according to colorectal cancer molecular subtypes based on gene expression. The tumor organoids displayed much greater molecular heterogeneity compared to the normal tissue-derived organoids. Differential gene expression analysis revealed the upregulation of colon cancer-associated genes such as PROX1, BAMBI, PTCH1, and the Wnt target gene APCDD1 in tumor organoids [34, 46]. Another key function of bulk RNA-seq is assessing gene expression profiles and their role in drug response pathways. In neuroendocrine neoplasms (NENs), bulk RNA-seq revealed a high correlation of gene expression between PDOs and their tissue of origin across multiple passages, providing insight into the molecular pathways that drive drug responses [23]. Similarly, in liver cancer PDOs, bulk RNA-seq confirmed matching expression data of specific tumor and differentiation markers between PDOs and their corresponding parental tissue [37]. Bulk RNA-seq can also be used to track the stability of gene expression profiles over time. In a longitudinal study on various cancer PDOs, gene expression was found to remain stable during passaging. When the organoids were treated with various agents, bulk RNA-seq showed that treated PDOs exhibited downregulated genes associated with metastasis, migration, and invasion [48]. Additionally, bulk RNA-seq is useful for understanding specific molecular changes in disease subtypes. In prostate cancer, bulk RNA-seq revealed that the most frequently upregulated genes in castration-resistant prostate cancer (e.g., TP53, FOXA1, PIK3R1, and SPOP) were consistent across organoid lines and tumor tissue. Androgen receptor (AR)-negative PDOs displayed similar gene expression patterns to known AR-negative castration-resistant tumors [38]. In breast cancer PDOs, it was observed that non-cancerous organoids exhibited considerably less heterogeneity compared to tumor organoids. Tumor PDOs were further subjected to gene expression-subtyping and were found to align with most expression-based breast cancer classifications [26, 49]. However, the authors noted the significant impact of the tumor microenvironment (TME) on cancer progression [50]. While bulk RNA-seq offers valuable insights, it does have limitations. Its inability to provide single-cell resolution means that cellular heterogeneity within tumors can obscure relevant molecular subpopulations. Additionally, bulk RNA-seq can allow highly expressed genes to dominate the signal, which may result in the underrepresentation of key regulatory pathways [51]. Despite these limitations, bulk RNA-seq remains a highly cost-effective and robust technology for assessing differential gene expression in PDOs, helping ensure that they recapitulate the molecular characteristics of their tissue of origin.

11.3.2 Single-Cell RNA Sequencing

Single-cell sequencing technology holds significant potential in identifying rare populations of cells that may be implicated in resistance and relapse during cancer treatment. The first step in this sequencing process is single-cell isolation. Several methods are used, including limiting dilution, FACS, and laser capture microdissection, but droplet-based microfluidic encapsulation, a technology provided by platforms like 10x Genomics, is the most commonly used method [52]. Following isolation, RNA-seq libraries are generated by cell lysis, reverse transcription into first-strand cDNA using uniquely barcoded beads for each cell, second-strand synthesis, and amplification of the cDNA. The amplified cDNA is then used to create sequencing libraries, which are sequenced to obtain transcriptomic data, followed by computational analysis [53]. The primary function of single-cell RNA sequencing (scRNA-seq) is to enable high-resolution detection of gene expression at the single-cell level. Unlike single-cell DNA sequencing, which is not commonly used in organoid research as of 2025, scRNA-seq is increasingly popular due to its ability to accurately characterize tumor heterogeneity. This method reveals distinct cellular subpopulations and their functional states [54]. ScRNA-seq also allows for the classification of individual cells into published molecular subtypes and the dissection of cell states and differentiation across all present cell types [55]. Another key function of scRNA-seq is the comparison of cellular heterogeneity between PDOs and their parental tumors. In urothelial bladder cancer organoids, scRNA-seq was used to compare cellular heterogeneity and found that the fraction of epithelial cells was increased, while the fraction of the TME was decreased in PDOs [56]. In hepatoblastoma organoids, scRNA-seq demonstrated that expression patterns of hepatic markers, WNT-pathway-related factors, and epithelial-mesenchymal transition (EMT) markers remained consistent across cellular subpopulations in both PDOs and parental tissue over extended periods of organoid culture [42]. ScRNA-seq is also valuable for studying the role of the microenvironment and drug responses in organoids. In a head and neck cancer project, scRNA-seq was used to investigate the effect of the culture environment on the PDOs. Organoids seeded in collagen type-I expressed markers of poor differentiation (K14) and invasive head and neck cancers, demonstrating that PDOs can recapitulate 3-dimensional invasion [57]. Similarly, in colorectal cancer organoids, scRNA-seq was used as an orthogonal technique to assess transcriptome signatures, which were then correlated with cell signaling network states during cancer progression. In drug-screens, PDOs showed upregulation of epithelial markers and downregulation of EMT markers [58]. In sarcomatoid bladder cancer organoids, scRNA-seq was used to correlate transcriptomic changes at the cellular level with drug-induced phenotypic effects [59]. Finally, scRNA-seq contributes to the growing field of global reference maps for human organoids. Within the Human Cell Atlas (HCA) initiative, the "HCA|Organoid" group was formed to create an organoid single-cell data portal and establish a network for organoid research [60]. However, the main limitations of scRNA-seq include its high cost, complex computational analysis, large data burden, and the loss of spatial context in some methods.

11.3.3 Spatial Transcriptomics

One of the major challenges in single-cell isolation is liberating viable cells from tissues without inducing stress, cell death, or aggregation. To address this challenge, transcriptomics at single-cell resolution has been conducted with the added benefit of preserving spatial context. These methods, referred to as spatial transcriptomics, provide information about the position of each cell relative to its neighbors, which determines the signals (such as autocrine, paracrine, juxtacrine) to which cells are exposed. These signals can influence important functions such as cell development, tissue homeostasis, and immune interactions [61]. Spatial transcriptomics aims to quantify the transcripts of a gene at distinct spatial locations within a tissue, allowing for an integrated view of gene expression while maintaining spatial context. There are two broad categories of spatial transcriptomics methods: imaging-based technologies and sequencing-based methods [62]. Imaging-based technologies, such as in situ hybridization (ISH) and in situ sequencing (ISS), enable direct imaging of mRNA in tissues. ISH uses polymerization of mRNA transcripts to single-stranded probes with complementary sequences, while ISS directly sequences transcripts within a tissue section using sequencing ligation technology [63]. Sequencing-based methods of spatial transcriptomics preserve spatial information through direct capture and location recording via microdissection and microfluidics or through ligation to spatially barcoded probes in a microarray [62]. Due to the three-dimensional structure of PDOs, they can effectively recapitulate spatial heterogeneity, such as when cells at the core of an organoid undergo necrosis due to lack of oxygen and nutrients, cells at the periphery proliferate. This ability allows PDOs to mimic avascular tumors in vitro [64, 65]. Traditionally, spatial transcriptomics has been designed for tissue sections, which are typically two-dimensional constructs. However, it is more challenging to retain three-dimensional spatial information while obtaining high-content gene expression at single-cell resolution, especially in PDOs [66]. In cerebral organoids, imaging-based spatial transcriptomics has been applied by cryosectioning organoids into 10 μm tissue sections [67]. Similarly, spatial transcriptomics has been applied to PDOs derived from various tumor types, including melanoma and breast cancer. This approach involves suspending PDOs in a dye solution, allowing partial diffusion across the periphery-core axis. Single cells are then isolated and assessed for their dye content using fluorescence-activated cell sorting (FACS) and sequenced using a high-throughput scRNA-seq method, such as Smart-seq3, which retains full-length transcript coverage [68]. Another method uses photoactivatable cell dyes to retain spatial information on single-cell gene expression in intestinal organoids. PDOs are immersed in photoactivatable dyes, after which regions of interest are illuminated with a spatially controlled light source. The photoactivated regions can be tracked downstream during single-cell isolation and sequencing [69].

11.4 Epigenomics

Epigenetic mechanisms impact the stability of cellular processes, such as differentiation, cell-states, aging, and facilitate responses to environmental or developmental stimuli, without changing the underlying nucleic structure [70–72]. In addition, epigenetic aberrations play a profound role in tumorigenesis, progression, and response to treatments [9, 73, 74]. Epigenetic modifications encompass chromatin status, DNA methylation, and non-coding RNA-associated gene silencing, which determine the subsequent gene expression [75, 76]. Epigenomics enable us to understand the mechanisms of tumor heterogeneity and to identify potentially targetable variations as an approach to precision medicine [77]. This section will explore the most commonly used technologies in this still emerging field to assessing epigenetic changes and their relevance in organoid research.

11.4.1 Chromatin Variation

Several different factors influence chromatin architecture and histone location, which crucially determine the accessibility of the DNA sequence to induce transcription, repair, and replication [73]. One technology to examine chromatin-related epigenomic changes is single-cell assay for transposase-accessible chromatin using sequencing (scATAC-seq), which enables the assessment of chromatin accessibility for transcription factors at the individual cell level [76]. At single-cell resolution, scATAC-seq has become the most widely used assay to measure chromatin accessibility in recent years [52]. In hepatoblastoma-derived organoids, scATAC-seq was used to identify distinct enrichment patterns of transcription factor motifs, thus validating the presence of distinct regulatory networks driving molecular heterogeneity in tumor subtypes [42]. In colorectal cancer, scATAC-seq allowed for the comparison of epigenetic changes across primary tumors, PDOs, xenograft models (PDXs), and PDO-xenograft models (PDOXs). Notably, chromatin landscapes in PDOX were closer to PDX than to PDO, highlighting the influence of the TME on epigenetic adaptation [78]. In pediatric malignant rhabdoid tumors (MRT), scATAC-seq performed on PDOs revealed a novel epigenomic alteration as a mechanism of cancer development [79]. In liver cancer PDOs, scATAC-seq was used to assess epigenetic changes in response to a tyrosin kinase inhibitor (donafenib) and an epigenetic drug (GSK-J4), an inhibitor of histone demethylases [80]. In prostate cancer PDOs, PDXs, and cell lines, bulk ATAC-seq was used to identify regulatory regions associated with tumor subtypes based on chromatin accessibility [81]. Bulk ATAC-seq-based signatures were also developed in pancreatic cancer PDOs predictive of patient prognosis [82]. Another technology used to assess chromatin status is chromatin immunoprecipitation sequencing (ChIP-seq), which maps histone modifications and protein-DNA interactions across the genome to examine gene regulation. Compared to ATAC-seq, ChIP-seq provides targeted insights into specific protein-DNA interactions, requiring antibodies for known transcription factors or histone modifications, whereas ATAC-seq offers a broader, unbiased view of chromatin accessibility without prior knowledge of binding proteins [83]. For example, in hepatocellular carcinoma organoids, ChIP-seq was used to study how mutations reshape transcription factor binding and chromatin regulation, revealing gene expression changes linked to tumorigenesis [84]. In Lynch-Syndrome organoids, ChIP-seq was used to assess epigenetic changes of organoids during methyltransferase inhibition (EZH2 inhibition) as a potential prevention strategy for Lynch-syndrome patients, as EZH2 inhibition enhanced immune response and reduced adenoma formation in mouse models [85]. EZH2 inhibition was also examined in PDOs derived from lung squamous carcinoma to determine how chromatin states and gene expression are affected, to support its use as a therapeutic strategy to improve checkpoint inhibitor responses [86]. Another antibody-targeted technology for assessing epigenetic changes is Cleavage Under Targets and Tagmentation (CUT&Tag), which enables high-resolution profiling of histone modifications, transcription factors, and chromatin-associated proteins, including at the single-cell level [87]. CUT&Tag requires only a limited number of cells, allowing for high-resolution profiling of protein-DNA interactions, which makes it particularly useful for organoid research. In gastric cancer organoids, CUT&Tag identified binding sites associated with drug resistance [88]. In liver cancer organoids, CUT&Tag revealed that inhibiting TGM2 (transglutaminase 2) suppressed tumor growth [89]. To date, other technologies to examine chromatin changes, such as CUT&Run [90], spatial scATAC-seq [91], Hi-C [92], sortChic [93], or older technologies such as DNase-seq do not seem to be widely used in organoid studies.

11.4.2 DNA Methylation

As described, DNA methylation of cytosine is an epigenomic mechanism to regulate gene expression [94]. DNA-methylation sequencing data can be generated using various techniques such as bisulfite-sequencing [52]. Bisulfite sequencing is a method used to detect DNA methylation by treating DNA with bisulfite, which converts unmethylated cytosines to uracil while leaving methylated cytosines unchanged. After PCR amplification and sequencing, comparisons with the reference genome reveal the methylation status of individual cytosines at single-base resolution [95]. Bisulfite sequencing can be performed at different levels of resolution, including whole-genome bisulfite sequencing (WGBS) for comprehensive methylome analysis, reduced representation bisulfite sequencing (RRBS) for a targeted approach, and single-cell bisulfite sequencing (scBS-Seq) to assess DNA methylation at the single-cell level [96, 97]. In pancreatic cancer organoids, WGBS was performed on PDOs from different disease stages and demonstrated that PDOs of later-stage pancreatic cancers had lower global methylation than earlier stage PDOs, highlighting epigenetic reprogramming during cancer progression. Thus, DNA methylation patterns could serve as potential biomarkers of disease stage [98].

11.5 Discussion

Integrating next-generation sequencing technologies with organoid research has significantly advanced our understanding of cancer by providing detailed insights into the genetic, transcriptomic, and epigenomic landscape of tumors (Fig. 11.2). Organoids serve as accurate cancer models, enabling the identification of genetic mutations, the study of gene expression, and the exploration of epigenetic changes. Key techniques such as whole-genome sequencing, whole-exome sequencing, bulk RNA sequencing, and single-cell RNA sequencing have been instrumental in cancer research, while methods like chromatin accessibility assays and DNA methylation sequencing have uncovered important regulatory mechanisms in tumor development.

A summary of these technologies is provided in Table 11.1. Combining sequencing approaches—such as linking whole-genome or whole-exome sequencing with RNA sequencing or integrating spatial transcriptomics with single-cell RNA sequencing—offers a more comprehensive view of tumor biology. When applied together, these technologies provide deeper insights into how mutations influence gene expression, cellular differentiation, and regulatory networks. While this review does not cover proteomics and metabolomics, their inclusion in future multi-omics studies will be essential for a more complete understanding of cancer biology and treatment responses.

Despite these advances, challenges remain. Sample preparation, particularly for single-cell and spatial transcriptomics, is complex due to the difficulty of dissociating cells from the extracellular matrix. The high cost of sequencing, especially in large-scale studies, limits its accessibility, though bulk sequencing remains a more

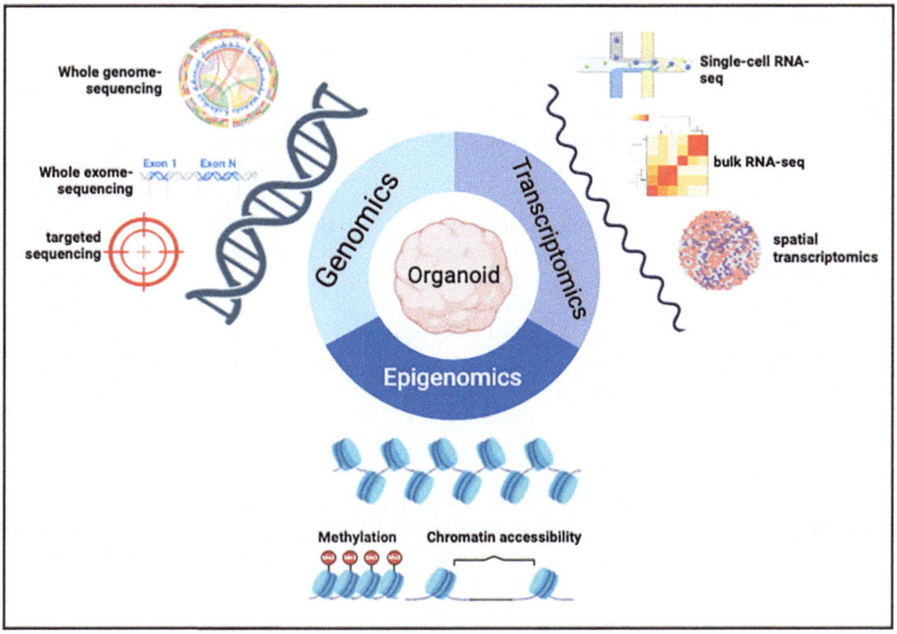

Fig. 11.2 Highlighted NGS methods in organoid research

Table 11.1 A summary of the sequencing technologies

Method	Resolution	Key advantages	Key limitations	Best for
WGS (Whole-Genome Sequencing)	Whole-genome	Comprehensive, detects mutations, SVs, CNAs, and fusions	Expensive, high data burden	Characterizing genomic instability, mutation profiling
WES (Whole-Exome Sequencing)	Exome-wide	Cost-effective, focuses on protein-coding genes	Misses regulatory regions, limited to exonic mutations	Studying disease-associated mutations, mutational analysis
Bulk RNA-seq	Population-level	Robust, cost-effective, detects differential gene expression	No single-cell resolution, loses heterogeneity	General transcriptomic profiling, gene expression in bulk
scRNA-seq (Single-cell RNA sequencing)	Single-cell	High resolution, reveals cellular heterogeneity, identifies rare populations	Costly, complex bioinformatics, high data burden	Identifying rare cell populations, lineage tracing, tumor heterogeneity

(continued)

Table 11.1 (continued)

Method	Resolution	Key advantages	Key limitations	Best for
Spatial Transcriptomics	Tissue context (single-cell)	Preserves spatial information of gene expression	Resolution limitations, tissue-specific variability	Studying cell-cell interactions, tissue organization, 3D spatial context
scATAC-seq (Chromatin Accessibility)	Single-cell	Measures chromatin accessibility, identifies regulatory regions	High cost, complex analysis	Studying chromatin states, gene regulation, tumor subtypes
ChIP-seq (Chromatin Immunoprecipitation)	Whole-genome	Maps histone modifications and protein-DNA interactions	Requires antibodies, targeted approach	Identifying specific protein-DNA interactions, gene regulation
CUT&Tag (Chromatin Profiling)	Single-cell	High-resolution, requires fewer cells, maps protein-DNA interactions	Limited to specific interactions, high complexity	Chromatin analysis, studying protein-DNA interactions at single-cell level
DNA methylation sequencing	Whole-genome	Measures DNA methylation at single-base resolution	High cost, limited to methylation patterns	Epigenetic regulation, tumorigenesis, disease stage biomarkers

affordable alternative. Additionally, integrating multi-omics data requires advanced bioinformatics tools and presents computational hurdles. Organoid models also exhibit variability, which can impact reproducibility across experiments, and standardized protocols for sequencing organoids are still evolving.

Nevertheless, continued advancements in sequencing technologies and organoid culture methods will further enhance their accuracy as cancer models, making them invaluable for both research and personalized medicine. Future efforts will focus on overcoming technical barriers, improving scalability, and refining multi-omics integration to deepen our understanding of cancer and develop more effective therapies.

11.6 Conclusion

The combination of next-generation sequencing technologies with organoid models has transformed cancer research. While organoids closely mimic the genetic, transcriptomic, and epigenomic features of primary tumors, it was the application of NGS that validated their fidelity and established them as powerful preclinical models. By enabling detailed genomic and molecular characterization, NGS has positioned organoids as essential tools for studying tumor biology and advancing personalized cancer therapies. Despite ongoing challenges in cost, sample preparation, and data integration, continued improvements are expected to enhance their scalability and utility in research and precision medicine.

References

1. Lancaster MA, Knoblich JA. Organogenesis in a dish: modeling development and disease using organoid technologies. Science. 2014;345:1247125. https://doi.org/10.1126/science.1247125.
2. Radić M, Egger M, Kruithof-de Julio M, Seiler R. Patient-derived organoids in bladder cancer: opportunities and challenges. Eur Urol Focus. 2024;11:62–70. https://doi.org/10.1016/j.euf.2024.08.008.
3. Kretzschmar K, Clevers H. Organoids: Modeling development and the stem cell niche in a dish. Dev Cell. 2016;38:590–600. https://doi.org/10.1016/j.devcel.2016.08.014.
4. McCauley HA, Wells JM. Pluripotent stem cell-derived organoids: using principles of developmental biology to grow human tissues in a dish. Development. 2017;144:958–62. https://doi.org/10.1242/dev.140731.
5. Clevers H. Modeling development and disease with organoids. Cell. 2016;165:1586–97. https://doi.org/10.1016/j.cell.2016.05.082.
6. Voest EE, Bernards R. DNA-guided precision medicine for cancer: a case of irrational exuberance? Cancer Discov. 2016;6:130–2. https://doi.org/10.1158/2159-8290.Cd-15-1321.
7. Zehir A, et al. Mutational landscape of metastatic cancer revealed from prospective clinical sequencing of 10,000 patients. Nat Med. 2017;23:703–13. https://doi.org/10.1038/nm.4333.
8. Pugh TJ, et al. The genetic landscape of high-risk neuroblastoma. Nat Genet. 2013;45:279–84. https://doi.org/10.1038/ng.2529.
9. Manolio TA, et al. Opportunities, resources, and techniques for implementing genomics in clinical care. Lancet. 2019;394:511–20. https://doi.org/10.1016/s0140-6736(19)31140-7.
10. Gerasimavicius L, Livesey BJ, Marsh JA. Loss-of-function, gain-of-function and dominant-negative mutations have profoundly different effects on protein structure. Nat Commun. 2022;13:3895. https://doi.org/10.1038/s41467-022-31686-6.
11. ENCODE Project Consortium. An integrated encyclopedia of DNA elements in the human genome. Nature. 2012;489:57–74. https://doi.org/10.1038/nature11247.
12. Vitsios D, Dhindsa RS, Middleton L, Gussow AB, Petrovski S. Prioritizing non-coding regions based on human genomic constraint and sequence context with deep learning. Nat Commun. 2021;12:1504. https://doi.org/10.1038/s41467-021-21790-4.
13. Makrythanasis P, Antonarakis SE. Pathogenic variants in non-protein-coding sequences. Clin Genet. 2013;84:422–8. https://doi.org/10.1111/cge.12272.
14. Ward LD, Kellis M. Interpreting noncoding genetic variation in complex traits and human disease. Nat Biotechnol. 2012;30:1095–106. https://doi.org/10.1038/nbt.2422.
15. Kinnersley B, et al. Analysis of 10,478 cancer genomes identifies candidate driver genes and opportunities for precision oncology. Nat Genet. 2024;56:1868–77. https://doi.org/10.1038/s41588-024-01785-9.
16. Pan-cancer analysis of whole genomes. Nature. 2020;578:82–93. https://doi.org/10.1038/s41586-020-1969-6.
17. Degasperi A, et al. Substitution mutational signatures in whole-genome-sequenced cancers in the UK population. Science. 2022;376:abl9283. https://doi.org/10.1126/science.abl9283.
18. Bagger FO, et al. Whole genome sequencing in clinical practice. BMC Med Genet. 2024;17:39. https://doi.org/10.1186/s12920-024-01795-w.
19. van Dijk EL, Jaszczyszyn Y, Naquin D, Thermes C. The third revolution in sequencing technology. Trends Genet. 2018;34:666–81. https://doi.org/10.1016/j.tig.2018.05.008.

20. Mizutani T, et al. Recapitulating the adenoma-carcinoma sequence by selection of four spontaneous oncogenic mutations in mismatch-repair-deficient human colon organoids. Nat Cancer. 2024;5:1852–67. https://doi.org/10.1038/s43018-024-00841-x.
21. Sachs N, et al. Long-term expanding human airway organoids for disease modeling. EMBO J. 2019;38:e100300. https://doi.org/10.15252/embj.2018100300.
22. Alcala N, et al. Multi-omic dataset of patient-derived tumor organoids of neuroendocrine neoplasms. Gigascience. 2024;13:giae008. https://doi.org/10.1093/gigascience/giae008.
23. Dayton TL, et al. Druggable growth dependencies and tumor evolution analysis in patient-derived organoids of neuroendocrine neoplasms from multiple body sites. Cancer Cell. 2023;41:2083–2099.e2089. https://doi.org/10.1016/j.ccell.2023.11.007.
24. Tiriac H, et al. Organoid profiling identifies common responders to chemotherapy in pancreatic cancer. Cancer Discov. 2018;8:1112–29. https://doi.org/10.1158/2159-8290.Cd-18-0349.
25. Huch M, et al. Long-term culture of genome-stable bipotent stem cells from adult human liver. Cell. 2015;160:299–312. https://doi.org/10.1016/j.cell.2014.11.050.
26. Sachs N, et al. A living biobank of breast cancer organoids captures disease heterogeneity. Cell. 2018;172:373–386.e310. https://doi.org/10.1016/j.cell.2017.11.010.
27. Youk J, Kwon HW, Kim R, Ju YS. Dissecting single-cell genomes through the clonal organoid technique. Exp Mol Med. 2021;53:1503–11. https://doi.org/10.1038/s12276-021-00680-1.
28. Molina-Ramírez LP, et al. Personalised virtual gene panels reduce interpretation workload and maintain diagnostic rates of proband-only clinical exome sequencing for rare disorders. J Med Genet. 2022;59:393–8. https://doi.org/10.1136/jmedgenet-2020-107303.
29. Méndez-Vidal C, et al. A genomic strategy for precision medicine in rare diseases: integrating customized algorithms into clinical practice. J Transl Med. 2025;23:86. https://doi.org/10.1186/s12967-025-06069-2.
30. Bamshad MJ, et al. Exome sequencing as a tool for mendelian disease gene discovery. Nat Rev Genet. 2011;12:745–55. https://doi.org/10.1038/nrg3031.
31. Majewski J, Schwartzentruber J, Lalonde E, Montpetit A, Jabado N. What can exome sequencing do for you? J Med Genet. 2011;48:580–9. https://doi.org/10.1136/jmedgenet-2011-100223.
32. Seaby EG, Pengelly RJ, Ennis S. Exome sequencing explained: a practical guide to its clinical application. Brief Funct Genomics. 2016;15:374–84. https://doi.org/10.1093/bfgp/elv054.
33. Seppälä TT, et al. Precision medicine in pancreatic cancer: patient-derived organoid Pharmacotyping is a predictive biomarker of clinical treatment response. Clin Cancer Res. 2022;28:3296–307. https://doi.org/10.1158/1078-0432.Ccr-21-4165.
34. van de Wetering M, et al. Prospective derivation of a living organoid biobank of colorectal cancer patients. Cell. 2015;161:933–45. https://doi.org/10.1016/j.cell.2015.03.053.
35. Millen R, et al. Patient-derived head and neck cancer organoids allow treatment stratification and serve as a tool for biomarker validation and identification. Med. 2023;4:290–310.e212. https://doi.org/10.1016/j.medj.2023.04.003.
36. Karthaus WR, et al. Identification of multipotent luminal progenitor cells in human prostate organoid cultures. Cell. 2014;159:163–75. https://doi.org/10.1016/j.cell.2014.08.017.
37. Broutier L, et al. Human primary liver cancer-derived organoid cultures for disease modeling and drug screening. Nat Med. 2017;23:1424–35. https://doi.org/10.1038/nm.4438.
38. Gao D, et al. Organoid cultures derived from patients with advanced prostate cancer. Cell. 2014;159:176–87. https://doi.org/10.1016/j.cell.2014.08.016.
39. Ballester LY, Luthra R, Kanagal-Shamanna R, Singh RR. Advances in clinical next-generation sequencing: target enrichment and sequencing technologies. Expert Rev Mol Diagn. 2016;16:357–72. https://doi.org/10.1586/14737159.2016.1133298.
40. Fisher KE, et al. Clinical validation and implementation of a targeted next-generation sequencing assay to detect somatic variants in non-small cell lung, melanoma, and gastrointestinal malignancies. J Mol Diagn. 2016;18:299–315. https://doi.org/10.1016/j.jmoldx.2015.11.006.
41. Saudi Mendeliome Group. Comprehensive gene panels provide advantages over clinical exome sequencing for Mendelian diseases. Genome Biol. 2015;16:134. https://doi.org/10.1186/s13059-015-0693-2.

42. Kluiver TA, et al. Divergent WNT signaling and drug sensitivity profiles within hepatoblastoma tumors and organoids. Nat Commun. 2024;15:8576. https://doi.org/10.1038/s41467-024-52757-w.
43. Vlachogiannis G, et al. Patient-derived organoids model treatment response of metastatic gastrointestinal cancers. Science. 2018;359:920–6. https://doi.org/10.1126/science.aao2774.
44. Golub TR, et al. Molecular classification of cancer: class discovery and class prediction by gene expression monitoring. Science. 1999;286:531–7. https://doi.org/10.1126/science.286.5439.531.
45. Joanito I, et al. Single-cell and bulk transcriptome sequencing identifies two epithelial tumor cell states and refines the consensus molecular classification of colorectal cancer. Nat Genet. 2022;54:963–75. https://doi.org/10.1038/s41588-022-01100-4.
46. Choi W, et al. Genetic alterations in the molecular subtypes of bladder cancer: illustration in the cancer genome atlas dataset. Eur Urol. 2017;72:354–65. https://doi.org/10.1016/j.eururo.2017.03.010.
47. Perou CM, et al. Molecular portraits of human breast tumours. Nature. 2000;406:747–52. https://doi.org/10.1038/35021093.
48. Xu X, et al. A living biobank of matched pairs of patient-derived xenografts and organoids for cancer pharmacology. PLoS One. 2023;18:e0279821. https://doi.org/10.1371/journal.pone.0279821.
49. Ali HR, et al. Genome-driven integrated classification of breast cancer validated in over 7,500 samples. Genome Biol. 2014;15:431. https://doi.org/10.1186/s13059-014-0431-1.
50. Soysal SD, Tzankov A, Muenst SE. Role of the tumor microenvironment in breast cancer. Pathobiology. 2015;82:142–52. https://doi.org/10.1159/000430499.
51. Trapnell C. Defining cell types and states with single-cell genomics. Genome Res. 2015;25:1491–8. https://doi.org/10.1101/gr.190595.115.
52. Lim B, Lin Y, Navin N. Advancing cancer research and medicine with single-cell genomics. Cancer Cell. 2020;37:456–70. https://doi.org/10.1016/j.ccell.2020.03.008.
53. Hwang B, Lee JH, Bang D. Single-cell RNA sequencing technologies and bioinformatics pipelines. Exp Mol Med. 2018;50:1–14. https://doi.org/10.1038/s12276-018-0071-8.
54. Kashima Y, et al. Single-cell sequencing techniques from individual to multiomics analyses. Exp Mol Med. 2020;52:1419–27. https://doi.org/10.1038/s12276-020-00499-2.
55. Camp JG, et al. Human cerebral organoids recapitulate gene expression programs of fetal neocortex development. Proc Natl Acad Sci USA. 2015;112:15672–7. https://doi.org/10.1073/pnas.1520760112.
56. Minoli M, et al. Bladder cancer organoids as a functional system to model different disease stages and therapy response. Nat Commun. 2023;14:2214. https://doi.org/10.1038/s41467-023-37696-2.
57. Haughton PD, et al. Differential transcriptional invasion signatures from patient derived organoid models define a functional prognostic tool for head and neck cancer. Oncogene. 2024;43:2463–74. https://doi.org/10.1038/s41388-024-03091-4.
58. Sell T, et al. Oncogenic signaling is coupled to colorectal cancer cell differentiation state. J Cell Biol. 2023;222:e202204001. https://doi.org/10.1083/jcb.202204001.
59. Garioni M, et al. Patient-derived organoids identify tailored therapeutic options and determinants of plasticity in sarcomatoid urothelial bladder cancer. NPJ Precis Oncol. 2023;7:112. https://doi.org/10.1038/s41698-023-00466-w.
60. Bock C, et al. The organoid cell atlas. Nat Biotechnol. 2021;39:13–7. https://doi.org/10.1038/s41587-020-00762-x.
61. Armingol E, Officer A, Harismendy O, Lewis NE. Deciphering cell-cell interactions and communication from gene expression. Nat Rev Genet. 2021;22:71–88. https://doi.org/10.1038/s41576-020-00292-x.
62. Williams CG, Lee HJ, Asatsuma T, Vento-Tormo R, Haque A. An introduction to spatial transcriptomics for biomedical research. Genome Med. 2022;14:68. https://doi.org/10.1186/s13073-022-01075-1.

63. Kuemmerle LB, et al. Probe set selection for targeted spatial transcriptomics. Nat Methods. 2024;21:2260–70. https://doi.org/10.1038/s41592-024-02496-z.
64. Sant S, Johnston PA. The production of 3D tumor spheroids for cancer drug discovery. Drug Discov Today Technol. 2017;23:27–36. https://doi.org/10.1016/j.ddtec.2017.03.002.
65. Cougnoux A, et al. Diffusion smart-seq3 of breast cancer spheroids to explore spatial tumor biology and test evolutionary principles of tumor heterogeneity. Sci Rep. 2025;15:3811. https://doi.org/10.1038/s41598-024-83989-x.
66. Wang X, et al. Three-dimensional intact-tissue sequencing of single-cell transcriptional states. Science. 2018;361:eaat5691. https://doi.org/10.1126/science.aat5691.
67. He Z, et al. Lineage recording in human cerebral organoids. Nat Methods. 2022;19:90–9. https://doi.org/10.1038/s41592-021-01344-8.
68. Hagemann-Jensen M, Ziegenhain C, Sandberg R. Scalable single-cell RNA sequencing from full transcripts with smart-seq3xpress. Nat Biotechnol. 2022;40:1452–7. https://doi.org/10.1038/s41587-022-01311-4.
69. Genshaft AS, et al. Live cell tagging tracking and isolation for spatial transcriptomics using photoactivatable cell dyes. Nat Commun. 2021;12:4995. https://doi.org/10.1038/s41467-021-25279-y.
70. Flavahan WA, Gaskell E, Bernstein BE. Epigenetic plasticity and the hallmarks of cancer. Science. 2017;357:eaal2380. https://doi.org/10.1126/science.aal2380.
71. Cavalli G, Heard E. Advances in epigenetics link genetics to the environment and disease. Nature. 2019;571:489–99. https://doi.org/10.1038/s41586-019-1411-0.
72. López-Otín C, Blasco MA, Partridge L, Serrano M, Kroemer G. Hallmarks of aging: an expanding universe. Cell. 2023;186:243–78. https://doi.org/10.1016/j.cell.2022.11.001.
73. Dawson MA. The cancer epigenome: concepts, challenges, and therapeutic opportunities. Science. 2017;355:1147–52. https://doi.org/10.1126/science.aam7304.
74. Nicoglou A, Merlin F. Epigenetics: a way to bridge the gap between biological fields. Stud Hist Phil Biol Biomed Sci. 2017;66:73–82. https://doi.org/10.1016/j.shpsc.2017.10.002.
75. Begum N, et al. Epigenetics in depression and gut-brain axis: a molecular crosstalk. Front Aging Neurosci. 2022;14:1048333. https://doi.org/10.3389/fnagi.2022.1048333.
76. Furlano K, et al. Epigenomics-guided precision oncology: chromatin variants in prostate tumor evolution. Int J Cancer. 2025. https://doi.org/10.1002/ijc.35327.
77. Gopi LK, Kidder BL. Integrative pan cancer analysis reveals epigenomic variation in cancer type and cell specific chromatin domains. Nat Commun. 2021;12:1419. https://doi.org/10.1038/s41467-021-21707-1.
78. Xiang K, et al. Chromatin remodeling in patient-derived colorectal cancer models. Adv Sci (Weinh). 2024;11:e2303379. https://doi.org/10.1002/advs.202303379.
79. Liu NQ, et al. SMARCB1 loss activates patient-specific distal oncogenic enhancers in malignant rhabdoid tumors. Nat Commun. 2023;14:7762. https://doi.org/10.1038/s41467-023-43498-3.
80. Zheng C, et al. Donafenib and GSK-J4 synergistically induce ferroptosis in liver cancer by upregulating HMOX1 expression. Adv Sci (Weinh). 2023;10:e2206798. https://doi.org/10.1002/advs.202206798.
81. Tang F, et al. Chromatin profiles classify castration-resistant prostate cancers suggesting therapeutic targets. Science. 2022;376:eabe1505. https://doi.org/10.1126/science.abe1505.
82. Dhara S, et al. Pancreatic cancer prognosis is predicted by an ATAC-array technology for assessing chromatin accessibility. Nat Commun. 2021;12:3044. https://doi.org/10.1038/s41467-021-23237-2.
83. Kidder BL, Hu G, Zhao K. ChIP-Seq: technical considerations for obtaining high-quality data. Nat Immunol. 2011;12:918–22. https://doi.org/10.1038/ni.2117.
84. Lam YK, et al. TP53 R249S mutation in hepatic organoids captures the predisposing cancer risk. Hepatology. 2023;78:727–40. https://doi.org/10.1002/hep.32802.
85. Bowen CM, et al. Inhibition of histone methyltransferase EZH2 for immune-interception of colorectal cancer in lynch syndrome. JCI Insight. 2025;10:e177545. https://doi.org/10.1172/jci.insight.177545.

86. DuCote TJ, et al. EZH2 inhibition promotes tumor immunogenicity in lung squamous cell carcinomas. bioRxiv. 2023. https://doi.org/10.1101/2023.06.06.543919.
87. Kaya-Okur HS, et al. CUT&tag for efficient epigenomic profiling of small samples and single cells. Nat Commun. 2019;10:1930. https://doi.org/10.1038/s41467-019-09982-5.
88. Yu J, et al. Super-enhancer activates master transcription factor NR3C1 expression and promotes 5-FU resistance in gastric cancer. Adv Sci (Weinh). 2025;12:e2409050. https://doi.org/10.1002/advs.202409050.
89. Dong R, et al. TGM2-mediated histone serotonylation promotes HCC progression via MYC signalling pathway. J Hepatol. 2025;83:105–18. https://doi.org/10.1016/j.jhep.2024.12.038.
90. Brahma S, Henikoff S. CUT&RUN profiling of the budding yeast epigenome. Methods Mol Biol. 2022;2477:129–47. https://doi.org/10.1007/978-1-0716-2257-5_9.
91. Llorens-Bobadilla E, et al. Solid-phase capture and profiling of open chromatin by spatial ATAC. Nat Biotechnol. 2023;41:1085–8. https://doi.org/10.1038/s41587-022-01603-9.
92. Ijaz J, et al. Haplotype-specific assembly of shattered chromosomes in esophageal adenocarcinomas. Cell Genom. 2024;4:100484. https://doi.org/10.1016/j.xgen.2023.100484.
93. Zeller P, et al. Single-cell sortChIC identifies hierarchical chromatin dynamics during hematopoiesis. Nat Genet. 2023;55:333–45. https://doi.org/10.1038/s41588-022-01260-3.
94. Zhao SG, et al. The DNA methylation landscape of advanced prostate cancer. Nat Genet. 2020;52:778–89. https://doi.org/10.1038/s41588-020-0648-8.
95. Liu P, Jacques J, Hwang CI. Epigenetic landscape of DNA methylation in pancreatic ductal adenocarcinoma. Epigenomes. 2024;8:41. https://doi.org/10.3390/epigenomes8040041.
96. Loyfer N, et al. A DNA methylation atlas of normal human cell types. Nature. 2023;613:355–64. https://doi.org/10.1038/s41586-022-05580-6.
97. Smallwood SA, et al. Single-cell genome-wide bisulfite sequencing for assessing epigenetic heterogeneity. Nat Methods. 2014;11:817–20. https://doi.org/10.1038/nmeth.3035.
98. Wang SS, et al. Whole-genome bisulfite sequencing identifies stage- and subtype-specific DNA methylation signatures in pancreatic cancer. iScience. 2024;27:109414. https://doi.org/10.1016/j.isci.2024.109414.

Index

O. S. Somuncu (ed.), *Organoids for Preclinical Disease Modeling and Personalized Medicine*, Stem Cell Biology and Regenerative Medicine 78,
https://doi.org/10.1007/978-3-032-10024-5

MIX
Papier aus verantwortungsvollen Quellen
Paper from responsible sources
FSC® C105338

If you have any concerns about our products,
you can contact us on
ProductSafety@springernature.com

In case Publisher is established outside the EU,
the EU authorized representative is:
Springer Nature Customer Service Center GmbH
Europaplatz 3, 69115 Heidelberg, Germany

Printed by Libri Plureos GmbH
in Hamburg, Germany